MICRODOSING

Tracker

"

Mushrooms are miniature pharmaceutical factories, and of the thousands of mushroom species in nature, our ancestors and modern scientists have identified several dozen that have a unique combination of talents that improve our health

PAUL STAMETS

Journal

One of the best practices when microdosing is to keep a journal to track your journey and find out what is working and what isn't working, so you can adjust. Remember it is advised to start small and adjust with small increases, dropping back the dosage weight when needed. I recommend you start journaling before you even start microdosing, so you can have a reference on how you were feeling before starting. And try to keep objective and be always honest.

Notes for the daily journal:

- Include the date of the last time you took it in "Last Intake".
- BOD stands for "Beginning of Day".
- EOD stands for "End of Day".
- In "Day" include both the day of the week and the day of the month.
- Fill in the drawn boxes with a scale that goes from one to ten, being zero null effect. You can do it later in the day but try to keep doing it at a regular time.
- The reflection/observations section can be completed at the end of the day, or perhaps the next day. It can include changes in your thoughts, behavior, and others.
- The daily benefit score goes from 1 to 100, being the total sum of the factors analyzed on that day.
- In "Form" you can include if the specimen is dry or fresh, as well as the potency or strain you are taking.

Notes for the monthly overview:

- The first graph is intended to illustrate the monthly trend of the overall benefit from your microdosing journey. You will need the daily score tracking for at least a whole month.
- The rest of the graphs can be personalized, you might want to illustrate your monthly trend of creativity, sleep quality, or calmness to reflect your anxiety levels. So, feel free to use it at your convenience.
- The graphs are made following the monthly timeframe, but again feel free to increase or decrease the counting days.

JOURNAL EXAMPLE

DAY/TIME	DOSE	FORM	LAST INTAKE
Sat 24th Dec, 9am	0.15 g	Dry Golden T. 0.63%	Thur 22th Dec, 7am

GOALS/INTENTIONS *I would love to stop fear from holding me back. I want to enjoy exercising this evening, as I am not very motivated lately...*

FEELINGS BOD *I feel like having more self-compassion for myself than in the previous weeks, I woke up at 8:45 am and I am not guilty of it.*

		NOTES
MOOD	9	
COGNITIVE FUNCTIONING	7	
CALMNESS	9	*I don't feel distressed anymore*
CONCENTRATION	7	
FOCUS / PRODUCTIVITY	7	*Less procrastination*
ENERGY LEVELS	6	
RELATIONAL SKILLS	8	*I felt very confident talking to my neighbors today.*
SENSES	7	*Was I seeing brighter colors? I appreciated more the music that's for sure.*
SLEEP	10	*I slept great and had some lucid dreams (I was in Africa).*
SENSE OF CONNECTION/ INTROSPECTION	9	*I loved today's hike; I felt a deep connection with Mother Earth*
DAILY BENEFIT SCORE	79	

JOURNAL EXAMPLE

My anxiety is gone.

I felt a little bit nervous this morning

I didn't have a lot of appetite for breakfast, perhaps I waited too long.

I feel strong and proud of myself. I also feel inspired.

I went exercising and it was raining, and I didn't care :) And I feel I am forgiving myself...

Monthly Overview Example

Overall Benefits Analyzer Example

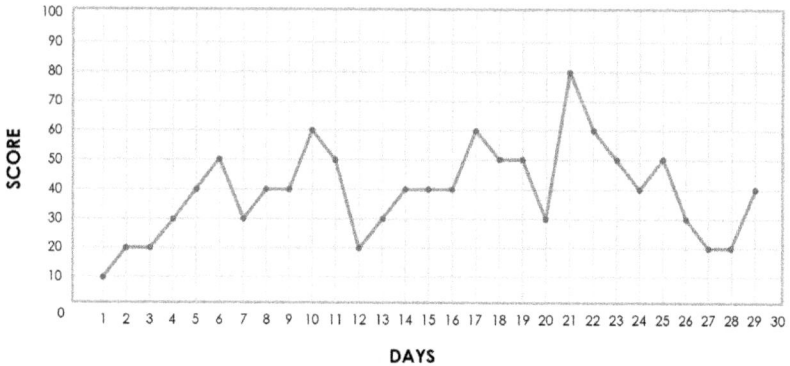

SCORE / DAYS

Mood Analyzer Example

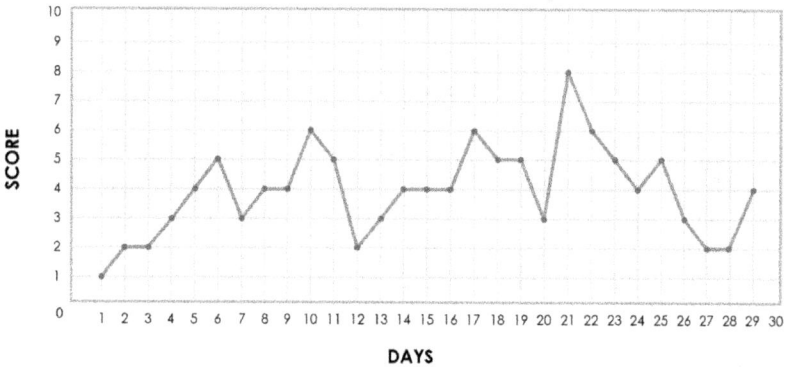

SCORE / DAYS

Clamness Analyzer (Inverted anxiety) Example

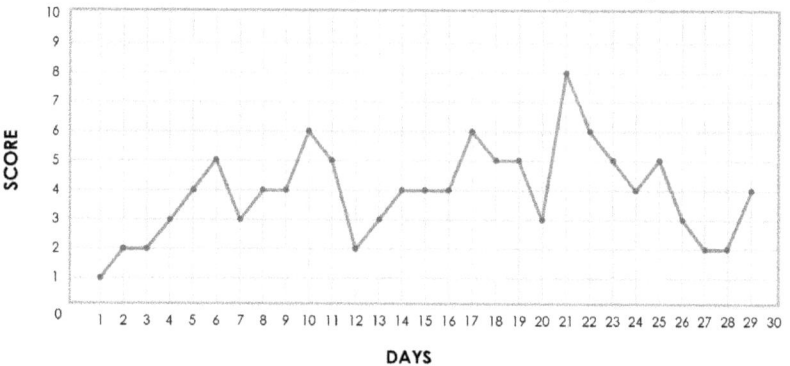

SCORE / DAYS

Before you get started, here are some related books you might want to check out:

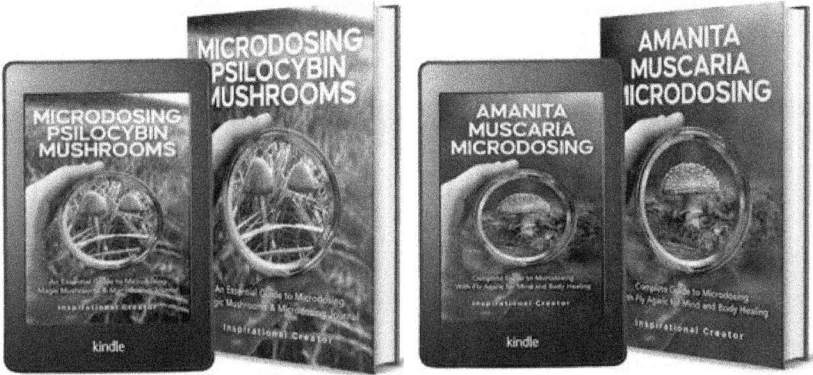

Microdosing Psilocybin Mushrooms: An Essential Guide to Microdosing Magic Mushrooms & Microdosing Journal by Bil Harret & Anastasia V. Sasha

"Amanita Muscaria Microdosing: Complete Guide to Microdosing With Fly Agaric for Mind and Body Healing, & Bonus" by Bil Harret & Anastasia V. Sasha

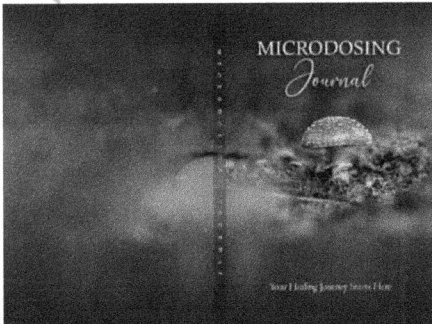

"Microdosing Journal: Amanita Muscaria (Fly agaric) Version. Your healing starts here"

DAY/TIME	DOSE	FORM	LAST INTAKE

GOALS/INTENTIONS

FEELINGS BOD

MOOD ☐ NOTES

COGNITIVE FUNCTIONING ☐ NOTES

CALMNESS ☐ NOTES

CONCENTRATION ☐ NOTES

FOCUS / PRODUCTIVITY ☐ NOTES

ENERGY LEVELS ☐ NOTES

RELATIONAL SKILLS ☐ NOTES

SENSES ☐ NOTES

SLEEP ☐ NOTES

SENSE OF CONNECTION/
INTROSPECTION ☐ NOTES

DAILY BENEFIT SCORE ☐ NOTES

EFFECTS ON HEALTH ISSUES

NEGATIVE EFFECTS

SIDE EFFECTS (AND SEVERITY)

FEELINGS EOD

REFLECTION/OBSERVATIONS

DAY/TIME	DOSE	FORM	LAST INTAKE

GOALS/INTENTIONS

FEELINGS BOD

MOOD ☐ NOTES

COGNITIVE FUNCTIONING ☐ NOTES

CALMNESS ☐ NOTES

CONCENTRATION ☐ NOTES

FOCUS / PRODUCTIVITY ☐ NOTES

ENERGY LEVELS ☐ NOTES

RELATIONAL SKILLS ☐ NOTES

SENSES ☐ NOTES

SLEEP ☐ NOTES

SENSE OF CONNECTION/
INTROSPECTION ☐ NOTES

DAILY BENEFIT SCORE ☐ NOTES

EFFECTS ON HEALTH ISSUES

NEGATIVE EFFECTS

SIDE EFFECTS (AND SEVERITY)

FEELINGS EOD

REFLECTION/OBSERVATIONS

DAY/TIME	DOSE	FORM	LAST INTAKE

GOALS/INTENTIONS

FEELINGS BOD

MOOD	☐	NOTES
COGNITIVE FUNCTIONING	☐	NOTES
CALMNESS	☐	NOTES
CONCENTRATION	☐	NOTES
FOCUS / PRODUCTIVITY	☐	NOTES
ENERGY LEVELS	☐	NOTES
RELATIONAL SKILLS	☐	NOTES
SENSES	☐	NOTES
SLEEP	☐	NOTES
SENSE OF CONNECTION/ INTROSPECTION	☐	NOTES
DAILY BENEFIT SCORE	☐	NOTES

EFFECTS ON HEALTH ISSUES

NEGATIVE EFFECTS

SIDE EFFECTS (AND SEVERITY)

FEELINGS EOD

REFLECTION/OBSERVATIONS

DAY/TIME	DOSE	FORM	LAST INTAKE

GOALS/INTENTIONS

FEELINGS BOD

MOOD ☐ NOTES

COGNITIVE FUNCTIONING ☐ NOTES

CALMNESS ☐ NOTES

CONCENTRATION ☐ NOTES

FOCUS / PRODUCTIVITY ☐ NOTES

ENERGY LEVELS ☐ NOTES

RELATIONAL SKILLS ☐ NOTES

SENSES ☐ NOTES

SLEEP ☐ NOTES

SENSE OF CONNECTION/ INTROSPECTION ☐ NOTES

DAILY BENEFIT SCORE ☐ NOTES

EFFECTS ON HEALTH ISSUES

NEGATIVE EFFECTS

SIDE EFFECTS (AND SEVERITY)

FEELINGS EOD

REFLECTION/OBSERVATIONS

DAY/TIME	DOSE	FORM	LAST INTAKE

GOALS/INTENTIONS

FEELINGS BOD

MOOD ☐ NOTES

COGNITIVE FUNCTIONING ☐ NOTES

CALMNESS ☐ NOTES

CONCENTRATION ☐ NOTES

FOCUS / PRODUCTIVITY ☐ NOTES

ENERGY LEVELS ☐ NOTES

RELATIONAL SKILLS ☐ NOTES

SENSES ☐ NOTES

SLEEP ☐ NOTES

SENSE OF CONNECTION/ INTROSPECTION ☐ NOTES

DAILY BENEFIT SCORE ☐ NOTES

EFFECTS ON HEALTH ISSUES

NEGATIVE EFFECTS

SIDE EFFECTS (AND SEVERITY)

FEELINGS EOD

REFLECTION/OBSERVATIONS

DAY/TIME	DOSE	FORM	LAST INTAKE

GOALS/INTENTIONS

FEELINGS BOD

MOOD ☐ NOTES

COGNITIVE FUNCTIONING ☐ NOTES

CALMNESS ☐ NOTES

CONCENTRATION ☐ NOTES

FOCUS / PRODUCTIVITY ☐ NOTES

ENERGY LEVELS ☐ NOTES

RELATIONAL SKILLS ☐ NOTES

SENSES ☐ NOTES

SLEEP ☐ NOTES

SENSE OF CONNECTION/ INTROSPECTION ☐ NOTES

DAILY BENEFIT SCORE ☐ NOTES

EFFECTS ON HEALTH ISSUES

NEGATIVE EFFECTS

SIDE EFFECTS (AND SEVERITY)

FEELINGS EOD

REFLECTION/OBSERVATIONS

DAY/TIME	DOSE	FORM	LAST INTAKE

GOALS/INTENTIONS

FEELINGS BOD

MOOD	☐	NOTES
COGNITIVE FUNCTIONING	☐	NOTES
CALMNESS	☐	NOTES
CONCENTRATION	☐	NOTES
FOCUS / PRODUCTIVITY	☐	NOTES
ENERGY LEVELS	☐	NOTES
RELATIONAL SKILLS	☐	NOTES
SENSES	☐	NOTES
SLEEP	☐	NOTES
SENSE OF CONNECTION/ INTROSPECTION	☐	NOTES
DAILY BENEFIT SCORE	☐	NOTES

EFFECTS ON HEALTH ISSUES

NEGATIVE EFFECTS

SIDE EFFECTS (AND SEVERITY)

FEELINGS EOD

REFLECTION/OBSERVATIONS

DAY/TIME	DOSE	FORM	LAST INTAKE

GOALS/INTENTIONS

FEELINGS BOD

MOOD ☐ NOTES

COGNITIVE FUNCTIONING ☐ NOTES

CALMNESS ☐ NOTES

CONCENTRATION ☐ NOTES

FOCUS / PRODUCTIVITY ☐ NOTES

ENERGY LEVELS ☐ NOTES

RELATIONAL SKILLS ☐ NOTES

SENSES ☐ NOTES

SLEEP ☐ NOTES

SENSE OF CONNECTION/
INTROSPECTION ☐ NOTES

DAILY BENEFIT SCORE ☐ NOTES

EFFECTS ON HEALTH ISSUES

NEGATIVE EFFECTS

SIDE EFFECTS (AND SEVERITY)

FEELINGS EOD

REFLECTION/OBSERVATIONS

DAY/TIME	DOSE	FORM	LAST INTAKE

GOALS/INTENTIONS

FEELINGS BOD

MOOD ☐ NOTES

COGNITIVE FUNCTIONING ☐ NOTES

CALMNESS ☐ NOTES

CONCENTRATION ☐ NOTES

FOCUS / PRODUCTIVITY ☐ NOTES

ENERGY LEVELS ☐ NOTES

RELATIONAL SKILLS ☐ NOTES

SENSES ☐ NOTES

SLEEP ☐ NOTES

SENSE OF CONNECTION/
INTROSPECTION ☐ NOTES

DAILY BENEFIT SCORE ☐ NOTES

EFFECTS ON HEALTH ISSUES

NEGATIVE EFFECTS

SIDE EFFECTS (AND SEVERITY)

FEELINGS EOD

REFLECTION/OBSERVATIONS

DAY/TIME	DOSE	FORM	LAST INTAKE

GOALS/INTENTIONS

FEELINGS BOD

MOOD ☐ NOTES

COGNITIVE FUNCTIONING ☐ NOTES

CALMNESS ☐ NOTES

CONCENTRATION ☐ NOTES

FOCUS / PRODUCTIVITY ☐ NOTES

ENERGY LEVELS ☐ NOTES

RELATIONAL SKILLS ☐ NOTES

SENSES ☐ NOTES

SLEEP ☐ NOTES

**SENSE OF CONNECTION/
INTROSPECTION** ☐ NOTES

DAILY BENEFIT SCORE ☐ NOTES

EFFECTS ON HEALTH ISSUES

NEGATIVE EFFECTS

SIDE EFFECTS (AND SEVERITY)

FEELINGS EOD

REFLECTION/OBSERVATIONS

DAY/TIME	DOSE	FORM	LAST INTAKE

GOALS/INTENTIONS

FEELINGS BOD

MOOD ☐ NOTES

COGNITIVE FUNCTIONING ☐ NOTES

CALMNESS ☐ NOTES

CONCENTRATION ☐ NOTES

FOCUS / PRODUCTIVITY ☐ NOTES

ENERGY LEVELS ☐ NOTES

RELATIONAL SKILLS ☐ NOTES

SENSES ☐ NOTES

SLEEP ☐ NOTES

SENSE OF CONNECTION/
INTROSPECTION ☐ NOTES

DAILY BENEFIT SCORE ☐ NOTES

EFFECTS ON HEALTH ISSUES

NEGATIVE EFFECTS

SIDE EFFECTS (AND SEVERITY)

FEELINGS EOD

REFLECTION/OBSERVATIONS

DAY/TIME	DOSE	FORM	LAST INTAKE

GOALS/INTENTIONS

FEELINGS BOD

MOOD ☐ NOTES

COGNITIVE FUNCTIONING ☐ NOTES

CALMNESS ☐ NOTES

CONCENTRATION ☐ NOTES

FOCUS / PRODUCTIVITY ☐ NOTES

ENERGY LEVELS ☐ NOTES

RELATIONAL SKILLS ☐ NOTES

SENSES ☐ NOTES

SLEEP ☐ NOTES

SENSE OF CONNECTION/ INTROSPECTION ☐ NOTES

DAILY BENEFIT SCORE ☐ NOTES

EFFECTS ON HEALTH ISSUES

NEGATIVE EFFECTS

SIDE EFFECTS (AND SEVERITY)

FEELINGS EOD

REFLECTION/OBSERVATIONS

DAY/TIME	DOSE	FORM	LAST INTAKE

GOALS/INTENTIONS

FEELINGS BOD

MOOD ☐ NOTES

COGNITIVE FUNCTIONING ☐ NOTES

CALMNESS ☐ NOTES

CONCENTRATION ☐ NOTES

FOCUS / PRODUCTIVITY ☐ NOTES

ENERGY LEVELS ☐ NOTES

RELATIONAL SKILLS ☐ NOTES

SENSES ☐ NOTES

SLEEP ☐ NOTES

SENSE OF CONNECTION/
INTROSPECTION ☐ NOTES

DAILY BENEFIT SCORE ☐ NOTES

EFFECTS ON HEALTH ISSUES

NEGATIVE EFFECTS

SIDE EFFECTS (AND SEVERITY)

FEELINGS EOD

REFLECTION/OBSERVATIONS

DAY/TIME	DOSE	FORM	LAST INTAKE

GOALS/INTENTIONS

FEELINGS BOD

MOOD ☐ NOTES

COGNITIVE FUNCTIONING ☐ NOTES

CALMNESS ☐ NOTES

CONCENTRATION ☐ NOTES

FOCUS / PRODUCTIVITY ☐ NOTES

ENERGY LEVELS ☐ NOTES

RELATIONAL SKILLS ☐ NOTES

SENSES ☐ NOTES

SLEEP ☐ NOTES

SENSE OF CONNECTION/
INTROSPECTION ☐ NOTES

DAILY BENEFIT SCORE ☐ NOTES

EFFECTS ON HEALTH ISSUES

NEGATIVE EFFECTS

SIDE EFFECTS (AND SEVERITY)

FEELINGS EOD

REFLECTION/OBSERVATIONS

DAY/TIME	DOSE	FORM	LAST INTAKE

GOALS/INTENTIONS

FEELINGS BOD

MOOD ☐ NOTES

COGNITIVE FUNCTIONING ☐ NOTES

CALMNESS ☐ NOTES

CONCENTRATION ☐ NOTES

FOCUS / PRODUCTIVITY ☐ NOTES

ENERGY LEVELS ☐ NOTES

RELATIONAL SKILLS ☐ NOTES

SENSES ☐ NOTES

SLEEP ☐ NOTES

SENSE OF CONNECTION/ INTROSPECTION ☐ NOTES

DAILY BENEFIT SCORE ☐ NOTES

EFFECTS ON HEALTH ISSUES

NEGATIVE EFFECTS

SIDE EFFECTS (AND SEVERITY)

FEELINGS EOD

REFLECTION/OBSERVATIONS

DAY/TIME	DOSE	FORM	LAST INTAKE

GOALS/INTENTIONS

FEELINGS BOD

MOOD	☐	NOTES
COGNITIVE FUNCTIONING	☐	NOTES
CALMNESS	☐	NOTES
CONCENTRATION	☐	NOTES
FOCUS / PRODUCTIVITY	☐	NOTES
ENERGY LEVELS	☐	NOTES
RELATIONAL SKILLS	☐	NOTES
SENSES	☐	NOTES
SLEEP	☐	NOTES
SENSE OF CONNECTION/ INTROSPECTION	☐	NOTES
DAILY BENEFIT SCORE	☐	NOTES

EFFECTS ON HEALTH ISSUES

NEGATIVE EFFECTS

SIDE EFFECTS (AND SEVERITY)

FEELINGS EOD

REFLECTION/OBSERVATIONS

DAY/TIME	DOSE	FORM	LAST INTAKE

GOALS/INTENTIONS

FEELINGS BOD

MOOD ☐ NOTES

COGNITIVE FUNCTIONING ☐ NOTES

CALMNESS ☐ NOTES

CONCENTRATION ☐ NOTES

FOCUS / PRODUCTIVITY ☐ NOTES

ENERGY LEVELS ☐ NOTES

RELATIONAL SKILLS ☐ NOTES

SENSES ☐ NOTES

SLEEP ☐ NOTES

SENSE OF CONNECTION/ INTROSPECTION ☐ NOTES

DAILY BENEFIT SCORE ☐ NOTES

EFFECTS ON HEALTH ISSUES

NEGATIVE EFFECTS

SIDE EFFECTS (AND SEVERITY)

FEELINGS EOD

REFLECTION/OBSERVATIONS

DAY/TIME	DOSE	FORM	LAST INTAKE

GOALS/INTENTIONS

FEELINGS BOD

MOOD ☐ NOTES

COGNITIVE FUNCTIONING ☐ NOTES

CALMNESS ☐ NOTES

CONCENTRATION ☐ NOTES

FOCUS / PRODUCTIVITY ☐ NOTES

ENERGY LEVELS ☐ NOTES

RELATIONAL SKILLS ☐ NOTES

SENSES ☐ NOTES

SLEEP ☐ NOTES

SENSE OF CONNECTION/ INTROSPECTION ☐ NOTES

DAILY BENEFIT SCORE ☐ NOTES

EFFECTS ON HEALTH ISSUES

NEGATIVE EFFECTS

SIDE EFFECTS (AND SEVERITY)

FEELINGS EOD

REFLECTION/OBSERVATIONS

DAY/TIME	DOSE	FORM	LAST INTAKE

GOALS/INTENTIONS

FEELINGS BOD

MOOD ☐ NOTES

COGNITIVE FUNCTIONING ☐ NOTES

CALMNESS ☐ NOTES

CONCENTRATION ☐ NOTES

FOCUS / PRODUCTIVITY ☐ NOTES

ENERGY LEVELS ☐ NOTES

RELATIONAL SKILLS ☐ NOTES

SENSES ☐ NOTES

SLEEP ☐ NOTES

SENSE OF CONNECTION/
INTROSPECTION ☐ NOTES

DAILY BENEFIT SCORE ☐ NOTES

EFFECTS ON HEALTH ISSUES

NEGATIVE EFFECTS

SIDE EFFECTS (AND SEVERITY)

FEELINGS EOD

REFLECTION/OBSERVATIONS

DAY/TIME	DOSE	FORM	LAST INTAKE

GOALS/INTENTIONS

FEELINGS BOD

MOOD ☐ NOTES

COGNITIVE FUNCTIONING ☐ NOTES

CALMNESS ☐ NOTES

CONCENTRATION ☐ NOTES

FOCUS / PRODUCTIVITY ☐ NOTES

ENERGY LEVELS ☐ NOTES

RELATIONAL SKILLS ☐ NOTES

SENSES ☐ NOTES

SLEEP ☐ NOTES

SENSE OF CONNECTION/
INTROSPECTION ☐ NOTES

DAILY BENEFIT SCORE ☐ NOTES

EFFECTS ON HEALTH ISSUES

NEGATIVE EFFECTS

SIDE EFFECTS (AND SEVERITY)

FEELINGS EOD

REFLECTION/OBSERVATIONS

DAY/TIME	DOSE	FORM	LAST INTAKE

GOALS/INTENTIONS

FEELINGS BOD

MOOD ☐ NOTES

COGNITIVE FUNCTIONING ☐ NOTES

CALMNESS ☐ NOTES

CONCENTRATION ☐ NOTES

FOCUS / PRODUCTIVITY ☐ NOTES

ENERGY LEVELS ☐ NOTES

RELATIONAL SKILLS ☐ NOTES

SENSES ☐ NOTES

SLEEP ☐ NOTES

SENSE OF CONNECTION/ INTROSPECTION ☐ NOTES

DAILY BENEFIT SCORE ☐ NOTES

EFFECTS ON HEALTH ISSUES

NEGATIVE EFFECTS

SIDE EFFECTS (AND SEVERITY)

FEELINGS EOD

REFLECTION/OBSERVATIONS

DAY/TIME	DOSE	FORM	LAST INTAKE

GOALS/INTENTIONS

FEELINGS BOD

MOOD	☐	NOTES
COGNITIVE FUNCTIONING	☐	NOTES
CALMNESS	☐	NOTES
CONCENTRATION	☐	NOTES
FOCUS / PRODUCTIVITY	☐	NOTES
ENERGY LEVELS	☐	NOTES
RELATIONAL SKILLS	☐	NOTES
SENSES	☐	NOTES
SLEEP	☐	NOTES
SENSE OF CONNECTION/ INTROSPECTION	☐	NOTES
DAILY BENEFIT SCORE	☐	NOTES

EFFECTS ON HEALTH ISSUES

NEGATIVE EFFECTS

SIDE EFFECTS (AND SEVERITY)

FEELINGS EOD

REFLECTION/OBSERVATIONS

DAY/TIME	DOSE	FORM	LAST INTAKE

GOALS/INTENTIONS

FEELINGS BOD

MOOD ☐ NOTES

COGNITIVE FUNCTIONING ☐ NOTES

CALMNESS ☐ NOTES

CONCENTRATION ☐ NOTES

FOCUS / PRODUCTIVITY ☐ NOTES

ENERGY LEVELS ☐ NOTES

RELATIONAL SKILLS ☐ NOTES

SENSES ☐ NOTES

SLEEP ☐ NOTES

SENSE OF CONNECTION/
INTROSPECTION ☐ NOTES

DAILY BENEFIT SCORE ☐ NOTES

EFFECTS ON HEALTH ISSUES

NEGATIVE EFFECTS

SIDE EFFECTS (AND SEVERITY)

FEELINGS EOD

REFLECTION/OBSERVATIONS

DAY/TIME	DOSE	FORM	LAST INTAKE

GOALS/INTENTIONS

FEELINGS BOD

MOOD ☐ NOTES

COGNITIVE FUNCTIONING ☐ NOTES

CALMNESS ☐ NOTES

CONCENTRATION ☐ NOTES

FOCUS / PRODUCTIVITY ☐ NOTES

ENERGY LEVELS ☐ NOTES

RELATIONAL SKILLS ☐ NOTES

SENSES ☐ NOTES

SLEEP ☐ NOTES

SENSE OF CONNECTION/ INTROSPECTION ☐ NOTES

DAILY BENEFIT SCORE ☐ NOTES

EFFECTS ON HEALTH ISSUES

NEGATIVE EFFECTS

SIDE EFFECTS (AND SEVERITY)

FEELINGS EOD

REFLECTION/OBSERVATIONS

DAY/TIME	DOSE	FORM	LAST INTAKE

GOALS/INTENTIONS

FEELINGS BOD

MOOD ☐ NOTES

COGNITIVE FUNCTIONING ☐ NOTES

CALMNESS ☐ NOTES

CONCENTRATION ☐ NOTES

FOCUS / PRODUCTIVITY ☐ NOTES

ENERGY LEVELS ☐ NOTES

RELATIONAL SKILLS ☐ NOTES

SENSES ☐ NOTES

SLEEP ☐ NOTES

SENSE OF CONNECTION/ INTROSPECTION ☐ NOTES

DAILY BENEFIT SCORE ☐ NOTES

EFFECTS ON HEALTH ISSUES

NEGATIVE EFFECTS

SIDE EFFECTS (AND SEVERITY)

FEELINGS EOD

REFLECTION/OBSERVATIONS

DAY/TIME	DOSE	FORM	LAST INTAKE

GOALS/INTENTIONS

FEELINGS BOD

MOOD	☐	NOTES
COGNITIVE FUNCTIONING	☐	NOTES
CALMNESS	☐	NOTES
CONCENTRATION	☐	NOTES
FOCUS / PRODUCTIVITY	☐	NOTES
ENERGY LEVELS	☐	NOTES
RELATIONAL SKILLS	☐	NOTES
SENSES	☐	NOTES
SLEEP	☐	NOTES
SENSE OF CONNECTION/ INTROSPECTION	☐	NOTES
DAILY BENEFIT SCORE	☐	NOTES

EFFECTS ON HEALTH ISSUES

NEGATIVE EFFECTS

SIDE EFFECTS (AND SEVERITY)

FEELINGS EOD

REFLECTION/OBSERVATIONS

DAY/TIME	DOSE	FORM	LAST INTAKE

GOALS/INTENTIONS

FEELINGS BOD

MOOD	☐	NOTES
COGNITIVE FUNCTIONING	☐	NOTES
CALMNESS	☐	NOTES
CONCENTRATION	☐	NOTES
FOCUS / PRODUCTIVITY	☐	NOTES
ENERGY LEVELS	☐	NOTES
RELATIONAL SKILLS	☐	NOTES
SENSES	☐	NOTES
SLEEP	☐	NOTES
SENSE OF CONNECTION/ INTROSPECTION	☐	NOTES
DAILY BENEFIT SCORE	☐	NOTES

EFFECTS ON HEALTH ISSUES

NEGATIVE EFFECTS

SIDE EFFECTS (AND SEVERITY)

FEELINGS EOD

REFLECTION/OBSERVATIONS

DAY/TIME	DOSE	FORM	LAST INTAKE

GOALS/INTENTIONS

FEELINGS BOD

MOOD ☐ NOTES

COGNITIVE FUNCTIONING ☐ NOTES

CALMNESS ☐ NOTES

CONCENTRATION ☐ NOTES

FOCUS / PRODUCTIVITY ☐ NOTES

ENERGY LEVELS ☐ NOTES

RELATIONAL SKILLS ☐ NOTES

SENSES ☐ NOTES

SLEEP ☐ NOTES

SENSE OF CONNECTION/ INTROSPECTION ☐ NOTES

DAILY BENEFIT SCORE ☐ NOTES

EFFECTS ON HEALTH ISSUES

NEGATIVE EFFECTS

SIDE EFFECTS (AND SEVERITY)

FEELINGS EOD

REFLECTION/OBSERVATIONS

DAY/TIME	DOSE	FORM	LAST INTAKE

GOALS/INTENTIONS

FEELINGS BOD

MOOD ☐ NOTES

COGNITIVE FUNCTIONING ☐ NOTES

CALMNESS ☐ NOTES

CONCENTRATION ☐ NOTES

FOCUS / PRODUCTIVITY ☐ NOTES

ENERGY LEVELS ☐ NOTES

RELATIONAL SKILLS ☐ NOTES

SENSES ☐ NOTES

SLEEP ☐ NOTES

SENSE OF CONNECTION/ INTROSPECTION ☐ NOTES

DAILY BENEFIT SCORE ☐ NOTES

EFFECTS ON HEALTH ISSUES

NEGATIVE EFFECTS

SIDE EFFECTS (AND SEVERITY)

FEELINGS EOD

REFLECTION/OBSERVATIONS

DAY/TIME	DOSE	FORM	LAST INTAKE

GOALS/INTENTIONS

FEELINGS BOD

		NOTES
MOOD	☐	
COGNITIVE FUNCTIONING	☐	
CALMNESS	☐	
CONCENTRATION	☐	
FOCUS / PRODUCTIVITY	☐	
ENERGY LEVELS	☐	
RELATIONAL SKILLS	☐	
SENSES	☐	
SLEEP	☐	
SENSE OF CONNECTION/ INTROSPECTION	☐	
DAILY BENEFIT SCORE	☐	

EFFECTS ON HEALTH ISSUES

NEGATIVE EFFECTS

SIDE EFFECTS (AND SEVERITY)

FEELINGS EOD

REFLECTION/OBSERVATIONS

Monthly Overview

Overall Benefits Analyzer

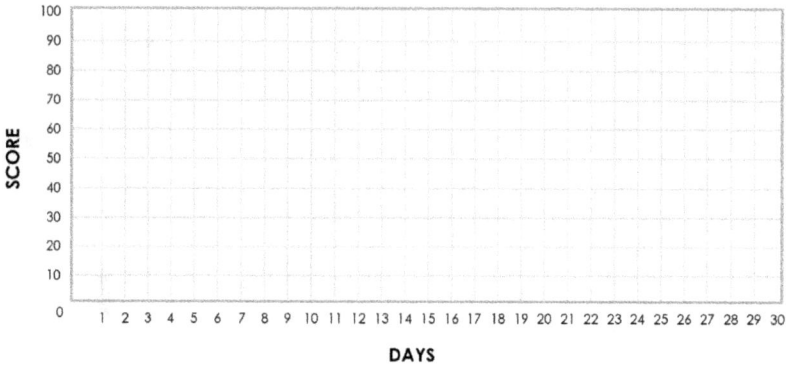

_____ Analyzer

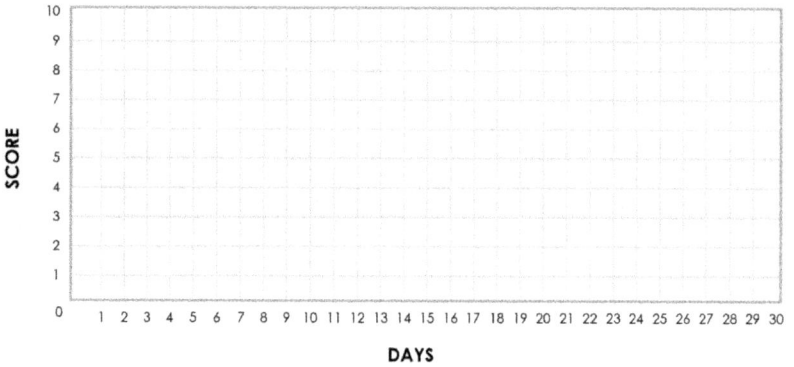

_____ Analyzer

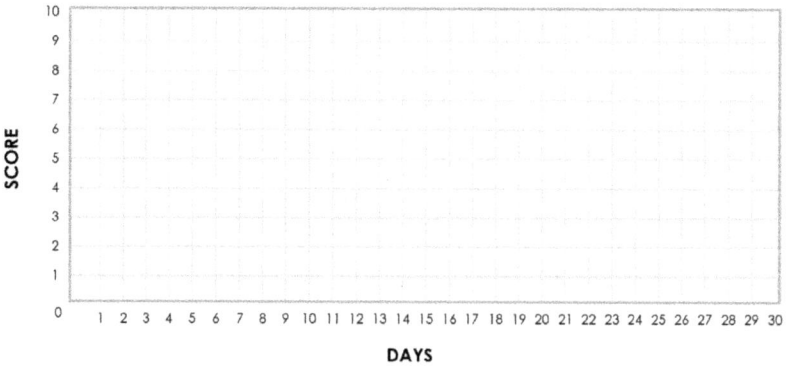

Monthly Overview

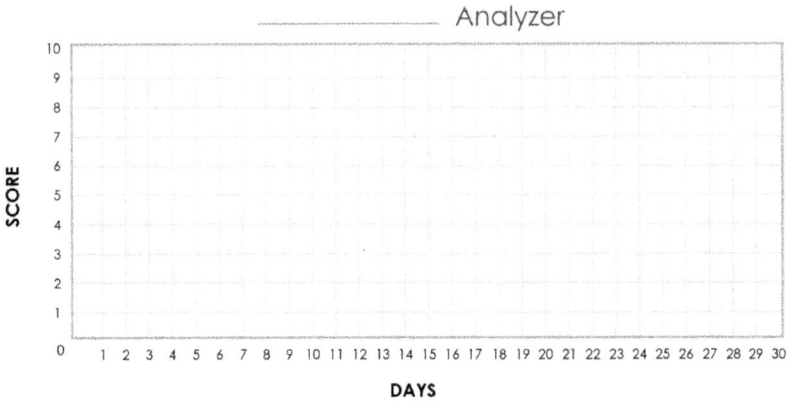

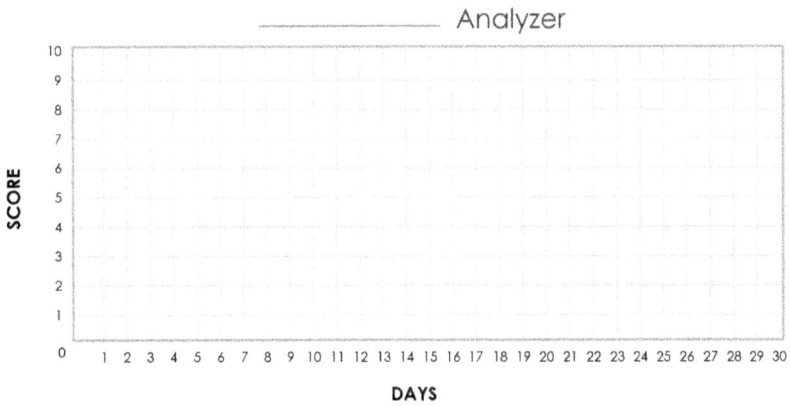

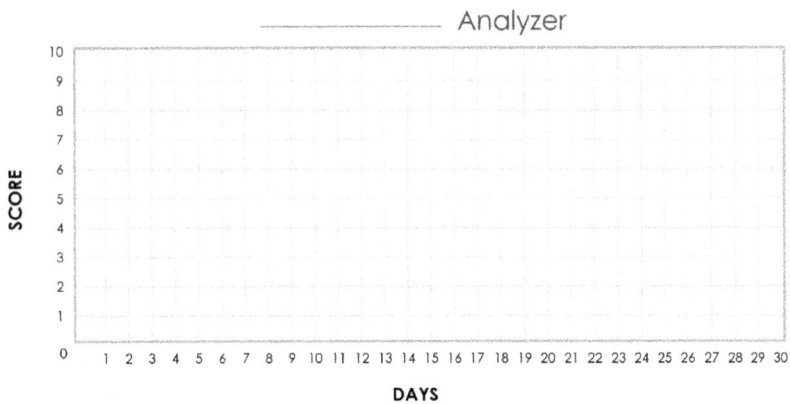

_____ Analyzer

SCORE

10
9
8
7
6
5
4
3
2
1
0

1 2 3 4 5 6 7 8 9 10 11 12 13 14 15 16 17 18 19 20 21 22 23 24 25 26 27 28 29 30

DAYS

_____ Analyzer

SCORE

10
9
8
7
6
5
4
3
2
1
0

1 2 3 4 5 6 7 8 9 10 11 12 13 14 15 16 17 18 19 20 21 22 23 24 25 26 27 28 29 30

DAYS

_____ Analyzer

SCORE

10
9
8
7
6
5
4
3
2
1
0

1 2 3 4 5 6 7 8 9 10 11 12 13 14 15 16 17 18 19 20 21 22 23 24 25 26 27 28 29 30

DAYS

Monthly Overview

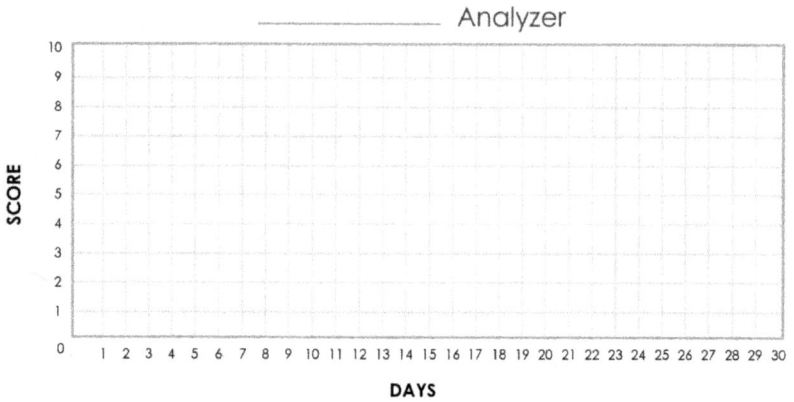

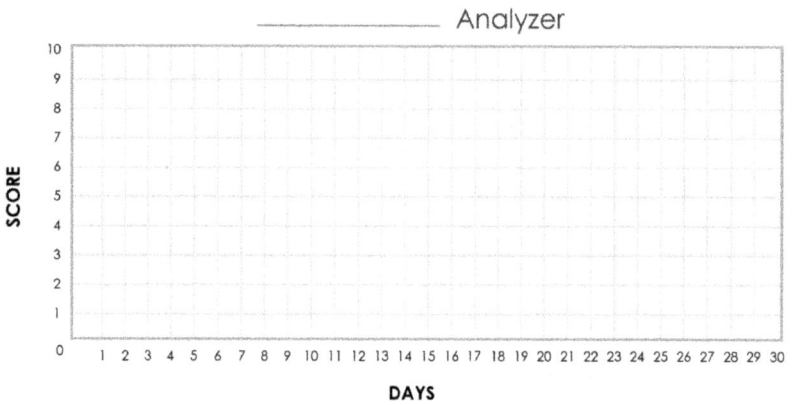

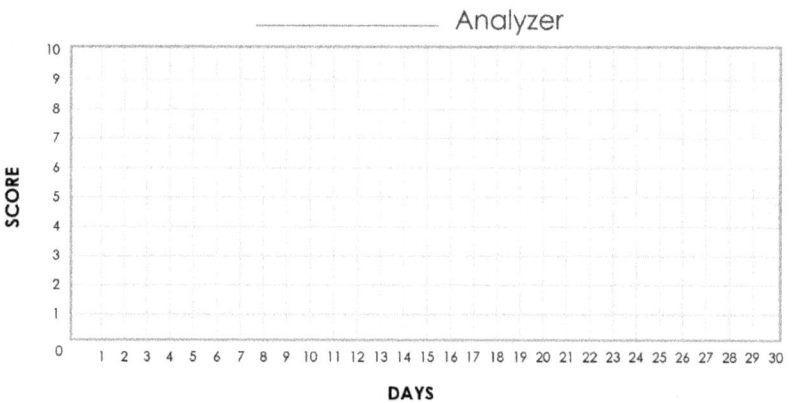

——————— Analyzer

SCORE

10
9
8
7
6
5
4
3
2
1
0

1 2 3 4 5 6 7 8 9 10 11 12 13 14 15 16 17 18 19 20 21 22 23 24 25 26 27 28 29 30

DAYS

——————— Analyzer

SCORE

10
9
8
7
6
5
4
3
2
1
0

1 2 3 4 5 6 7 8 9 10 11 12 13 14 15 16 17 18 19 20 21 22 23 24 25 26 27 28 29 30

DAYS

——————— Analyzer

SCORE

10
9
8
7
6
5
4
3
2
1
0

1 2 3 4 5 6 7 8 9 10 11 12 13 14 15 16 17 18 19 20 21 22 23 24 25 26 27 28 29 30

DAYS

Monthly Overview

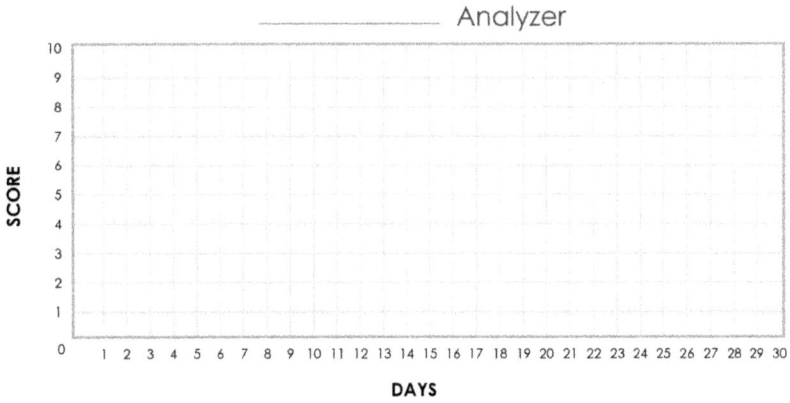

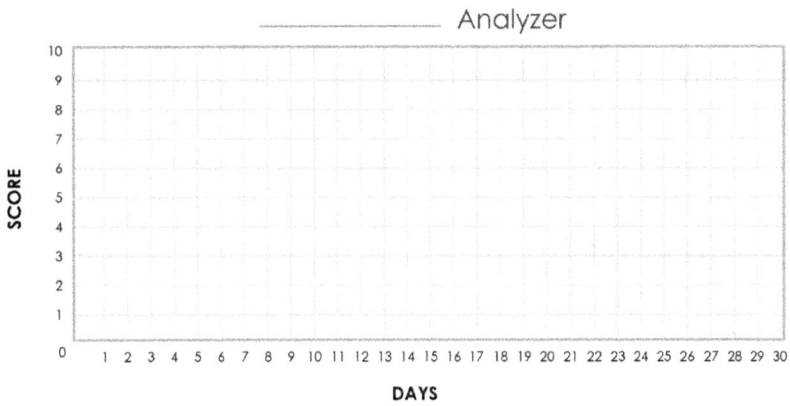

_____ Analyzer

SCORE

10
9
8
7
6
5
4
3
2
1
0

1 2 3 4 5 6 7 8 9 10 11 12 13 14 15 16 17 18 19 20 21 22 23 24 25 26 27 28 29 30

DAYS

_____ Analyzer

SCORE

10
9
8
7
6
5
4
3
2
1
0

1 2 3 4 5 6 7 8 9 10 11 12 13 14 15 16 17 18 19 20 21 22 23 24 25 26 27 28 29 30

DAYS

DAY/TIME	DOSE	FORM	LAST INTAKE

GOALS/INTENTIONS

FEELINGS BOD

MOOD	☐	NOTES
COGNITIVE FUNCTIONING	☐	NOTES
CALMNESS	☐	NOTES
CONCENTRATION	☐	NOTES
FOCUS / PRODUCTIVITY	☐	NOTES
ENERGY LEVELS	☐	NOTES
RELATIONAL SKILLS	☐	NOTES
SENSES	☐	NOTES
SLEEP	☐	NOTES
SENSE OF CONNECTION/ INTROSPECTION	☐	NOTES
DAILY BENEFIT SCORE	☐	NOTES

EFFECTS ON HEALTH ISSUES

NEGATIVE EFFECTS

SIDE EFFECTS (AND SEVERITY)

FEELINGS EOD

REFLECTION/OBSERVATIONS

DAY/TIME	DOSE	FORM	LAST INTAKE

GOALS/INTENTIONS

FEELINGS BOD

MOOD	☐	NOTES
COGNITIVE FUNCTIONING	☐	NOTES
CALMNESS	☐	NOTES
CONCENTRATION	☐	NOTES
FOCUS / PRODUCTIVITY	☐	NOTES
ENERGY LEVELS	☐	NOTES
RELATIONAL SKILLS	☐	NOTES
SENSES	☐	NOTES
SLEEP	☐	NOTES
SENSE OF CONNECTION/ INTROSPECTION	☐	NOTES
DAILY BENEFIT SCORE	☐	NOTES

EFFECTS ON HEALTH ISSUES

NEGATIVE EFFECTS

SIDE EFFECTS (AND SEVERITY)

FEELINGS EOD

REFLECTION/OBSERVATIONS

DAY/TIME	DOSE	FORM	LAST INTAKE

GOALS/INTENTIONS

FEELINGS BOD

MOOD ☐ NOTES

COGNITIVE FUNCTIONING ☐ NOTES

CALMNESS ☐ NOTES

CONCENTRATION ☐ NOTES

FOCUS / PRODUCTIVITY ☐ NOTES

ENERGY LEVELS ☐ NOTES

RELATIONAL SKILLS ☐ NOTES

SENSES ☐ NOTES

SLEEP ☐ NOTES

SENSE OF CONNECTION/ INTROSPECTION ☐ NOTES

DAILY BENEFIT SCORE ☐ NOTES

EFFECTS ON HEALTH ISSUES

NEGATIVE EFFECTS

SIDE EFFECTS (AND SEVERITY)

FEELINGS EOD

REFLECTION/OBSERVATIONS

DAY/TIME	DOSE	FORM	LAST INTAKE

GOALS/INTENTIONS

FEELINGS BOD

MOOD ☐ NOTES

COGNITIVE FUNCTIONING ☐ NOTES

CALMNESS ☐ NOTES

CONCENTRATION ☐ NOTES

FOCUS / PRODUCTIVITY ☐ NOTES

ENERGY LEVELS ☐ NOTES

RELATIONAL SKILLS ☐ NOTES

SENSES ☐ NOTES

SLEEP ☐ NOTES

SENSE OF CONNECTION/
INTROSPECTION ☐ NOTES

DAILY BENEFIT SCORE ☐ NOTES

EFFECTS ON HEALTH ISSUES

NEGATIVE EFFECTS

SIDE EFFECTS (AND SEVERITY)

FEELINGS EOD

REFLECTION/OBSERVATIONS

DAY/TIME	DOSE	FORM	LAST INTAKE

GOALS/INTENTIONS

FEELINGS BOD

MOOD ☐ NOTES

COGNITIVE FUNCTIONING ☐ NOTES

CALMNESS ☐ NOTES

CONCENTRATION ☐ NOTES

FOCUS / PRODUCTIVITY ☐ NOTES

ENERGY LEVELS ☐ NOTES

RELATIONAL SKILLS ☐ NOTES

SENSES ☐ NOTES

SLEEP ☐ NOTES

SENSE OF CONNECTION/ INTROSPECTION ☐ NOTES

DAILY BENEFIT SCORE ☐ NOTES

EFFECTS ON HEALTH ISSUES

NEGATIVE EFFECTS

SIDE EFFECTS (AND SEVERITY)

FEELINGS EOD

REFLECTION/OBSERVATIONS

DAY/TIME	DOSE	FORM	LAST INTAKE

GOALS/INTENTIONS

FEELINGS BOD

MOOD ☐ NOTES

COGNITIVE FUNCTIONING ☐ NOTES

CALMNESS ☐ NOTES

CONCENTRATION ☐ NOTES

FOCUS / PRODUCTIVITY ☐ NOTES

ENERGY LEVELS ☐ NOTES

RELATIONAL SKILLS ☐ NOTES

SENSES ☐ NOTES

SLEEP ☐ NOTES

SENSE OF CONNECTION/ INTROSPECTION ☐ NOTES

DAILY BENEFIT SCORE ☐ NOTES

EFFECTS ON HEALTH ISSUES

NEGATIVE EFFECTS

SIDE EFFECTS (AND SEVERITY)

FEELINGS EOD

REFLECTION/OBSERVATIONS

DAY/TIME	DOSE	FORM	LAST INTAKE

GOALS/INTENTIONS

FEELINGS BOD

		NOTES
MOOD	☐	NOTES
COGNITIVE FUNCTIONING	☐	NOTES
CALMNESS	☐	NOTES
CONCENTRATION	☐	NOTES
FOCUS / PRODUCTIVITY	☐	NOTES
ENERGY LEVELS	☐	NOTES
RELATIONAL SKILLS	☐	NOTES
SENSES	☐	NOTES
SLEEP	☐	NOTES
SENSE OF CONNECTION/ INTROSPECTION	☐	NOTES
DAILY BENEFIT SCORE	☐	NOTES

EFFECTS ON HEALTH ISSUES

NEGATIVE EFFECTS

SIDE EFFECTS (AND SEVERITY)

FEELINGS EOD

REFLECTION/OBSERVATIONS

DAY/TIME	DOSE	FORM	LAST INTAKE

GOALS/INTENTIONS

FEELINGS BOD

MOOD ☐ NOTES

COGNITIVE FUNCTIONING ☐ NOTES

CALMNESS ☐ NOTES

CONCENTRATION ☐ NOTES

FOCUS / PRODUCTIVITY ☐ NOTES

ENERGY LEVELS ☐ NOTES

RELATIONAL SKILLS ☐ NOTES

SENSES ☐ NOTES

SLEEP ☐ NOTES

SENSE OF CONNECTION/
INTROSPECTION ☐ NOTES

DAILY BENEFIT SCORE ☐ NOTES

EFFECTS ON HEALTH ISSUES

NEGATIVE EFFECTS

SIDE EFFECTS (AND SEVERITY)

FEELINGS EOD

REFLECTION/OBSERVATIONS

DAY/TIME	DOSE	FORM	LAST INTAKE

GOALS/INTENTIONS

FEELINGS BOD

MOOD ☐ NOTES

COGNITIVE FUNCTIONING ☐ NOTES

CALMNESS ☐ NOTES

CONCENTRATION ☐ NOTES

FOCUS / PRODUCTIVITY ☐ NOTES

ENERGY LEVELS ☐ NOTES

RELATIONAL SKILLS ☐ NOTES

SENSES ☐ NOTES

SLEEP ☐ NOTES

SENSE OF CONNECTION/
INTROSPECTION ☐ NOTES

DAILY BENEFIT SCORE ☐ NOTES

EFFECTS ON HEALTH ISSUES

NEGATIVE EFFECTS

SIDE EFFECTS (AND SEVERITY)

FEELINGS EOD

REFLECTION/OBSERVATIONS

DAY/TIME	DOSE	FORM	LAST INTAKE

GOALS/INTENTIONS

FEELINGS BOD

MOOD ☐ NOTES

COGNITIVE FUNCTIONING ☐ NOTES

CALMNESS ☐ NOTES

CONCENTRATION ☐ NOTES

FOCUS / PRODUCTIVITY ☐ NOTES

ENERGY LEVELS ☐ NOTES

RELATIONAL SKILLS ☐ NOTES

SENSES ☐ NOTES

SLEEP ☐ NOTES

SENSE OF CONNECTION/
INTROSPECTION ☐ NOTES

DAILY BENEFIT SCORE ☐ NOTES

EFFECTS ON HEALTH ISSUES

NEGATIVE EFFECTS

SIDE EFFECTS (AND SEVERITY)

FEELINGS EOD

REFLECTION/OBSERVATIONS

DAY/TIME	DOSE	FORM	LAST INTAKE

GOALS/INTENTIONS

FEELINGS BOD

MOOD	☐	NOTES
COGNITIVE FUNCTIONING	☐	NOTES
CALMNESS	☐	NOTES
CONCENTRATION	☐	NOTES
FOCUS / PRODUCTIVITY	☐	NOTES
ENERGY LEVELS	☐	NOTES
RELATIONAL SKILLS	☐	NOTES
SENSES	☐	NOTES
SLEEP	☐	NOTES
SENSE OF CONNECTION/ INTROSPECTION	☐	NOTES
DAILY BENEFIT SCORE	☐	NOTES

EFFECTS ON HEALTH ISSUES

NEGATIVE EFFECTS

SIDE EFFECTS (AND SEVERITY)

FEELINGS EOD

REFLECTION/OBSERVATIONS

DAY/TIME	DOSE	FORM	LAST INTAKE

GOALS/INTENTIONS

FEELINGS BOD

MOOD ☐ NOTES

COGNITIVE FUNCTIONING ☐ NOTES

CALMNESS ☐ NOTES

CONCENTRATION ☐ NOTES

FOCUS / PRODUCTIVITY ☐ NOTES

ENERGY LEVELS ☐ NOTES

RELATIONAL SKILLS ☐ NOTES

SENSES ☐ NOTES

SLEEP ☐ NOTES

SENSE OF CONNECTION/
INTROSPECTION ☐ NOTES

DAILY BENEFIT SCORE ☐ NOTES

EFFECTS ON HEALTH ISSUES

NEGATIVE EFFECTS

SIDE EFFECTS (AND SEVERITY)

FEELINGS EOD

REFLECTION/OBSERVATIONS

DAY/TIME	DOSE	FORM	LAST INTAKE

GOALS/INTENTIONS

FEELINGS BOD

MOOD ☐ NOTES

COGNITIVE FUNCTIONING ☐ NOTES

CALMNESS ☐ NOTES

CONCENTRATION ☐ NOTES

FOCUS / PRODUCTIVITY ☐ NOTES

ENERGY LEVELS ☐ NOTES

RELATIONAL SKILLS ☐ NOTES

SENSES ☐ NOTES

SLEEP ☐ NOTES

SENSE OF CONNECTION/
INTROSPECTION ☐ NOTES

DAILY BENEFIT SCORE ☐ NOTES

EFFECTS ON HEALTH ISSUES

NEGATIVE EFFECTS

SIDE EFFECTS (AND SEVERITY)

FEELINGS EOD

REFLECTION/OBSERVATIONS

DAY/TIME	DOSE	FORM	LAST INTAKE

GOALS/INTENTIONS

FEELINGS BOD

MOOD ☐ NOTES

COGNITIVE FUNCTIONING ☐ NOTES

CALMNESS ☐ NOTES

CONCENTRATION ☐ NOTES

FOCUS / PRODUCTIVITY ☐ NOTES

ENERGY LEVELS ☐ NOTES

RELATIONAL SKILLS ☐ NOTES

SENSES ☐ NOTES

SLEEP ☐ NOTES

SENSE OF CONNECTION/ INTROSPECTION ☐ NOTES

DAILY BENEFIT SCORE ☐ NOTES

EFFECTS ON HEALTH ISSUES

NEGATIVE EFFECTS

SIDE EFFECTS (AND SEVERITY)

FEELINGS EOD

REFLECTION/OBSERVATIONS

DAY/TIME	DOSE	FORM	LAST INTAKE

GOALS/INTENTIONS

FEELINGS BOD

MOOD ☐ NOTES

COGNITIVE FUNCTIONING ☐ NOTES

CALMNESS ☐ NOTES

CONCENTRATION ☐ NOTES

FOCUS / PRODUCTIVITY ☐ NOTES

ENERGY LEVELS ☐ NOTES

RELATIONAL SKILLS ☐ NOTES

SENSES ☐ NOTES

SLEEP ☐ NOTES

SENSE OF CONNECTION/ INTROSPECTION ☐ NOTES

DAILY BENEFIT SCORE ☐ NOTES

EFFECTS ON HEALTH ISSUES

NEGATIVE EFFECTS

SIDE EFFECTS (AND SEVERITY)

FEELINGS EOD

REFLECTION/OBSERVATIONS

DAY/TIME	DOSE	FORM	LAST INTAKE

GOALS/INTENTIONS

FEELINGS BOD

MOOD ☐ NOTES

COGNITIVE FUNCTIONING ☐ NOTES

CALMNESS ☐ NOTES

CONCENTRATION ☐ NOTES

FOCUS / PRODUCTIVITY ☐ NOTES

ENERGY LEVELS ☐ NOTES

RELATIONAL SKILLS ☐ NOTES

SENSES ☐ NOTES

SLEEP ☐ NOTES

SENSE OF CONNECTION/ INTROSPECTION ☐ NOTES

DAILY BENEFIT SCORE ☐ NOTES

EFFECTS ON HEALTH ISSUES

NEGATIVE EFFECTS

SIDE EFFECTS (AND SEVERITY)

FEELINGS EOD

REFLECTION/OBSERVATIONS

DAY/TIME	DOSE	FORM	LAST INTAKE

GOALS/INTENTIONS

FEELINGS BOD

MOOD ☐ NOTES

COGNITIVE FUNCTIONING ☐ NOTES

CALMNESS ☐ NOTES

CONCENTRATION ☐ NOTES

FOCUS / PRODUCTIVITY ☐ NOTES

ENERGY LEVELS ☐ NOTES

RELATIONAL SKILLS ☐ NOTES

SENSES ☐ NOTES

SLEEP ☐ NOTES

SENSE OF CONNECTION/ INTROSPECTION ☐ NOTES

DAILY BENEFIT SCORE ☐ NOTES

EFFECTS ON HEALTH ISSUES

NEGATIVE EFFECTS

SIDE EFFECTS (AND SEVERITY)

FEELINGS EOD

REFLECTION/OBSERVATIONS

DAY/TIME	DOSE	FORM	LAST INTAKE

GOALS/INTENTIONS

FEELINGS BOD

MOOD ☐ NOTES

COGNITIVE FUNCTIONING ☐ NOTES

CALMNESS ☐ NOTES

CONCENTRATION ☐ NOTES

FOCUS / PRODUCTIVITY ☐ NOTES

ENERGY LEVELS ☐ NOTES

RELATIONAL SKILLS ☐ NOTES

SENSES ☐ NOTES

SLEEP ☐ NOTES

SENSE OF CONNECTION/
INTROSPECTION ☐ NOTES

DAILY BENEFIT SCORE ☐ NOTES

EFFECTS ON HEALTH ISSUES

NEGATIVE EFFECTS

SIDE EFFECTS (AND SEVERITY)

FEELINGS EOD

REFLECTION/OBSERVATIONS

DAY/TIME	DOSE	FORM	LAST INTAKE

GOALS/INTENTIONS

FEELINGS BOD

MOOD ☐ NOTES

COGNITIVE FUNCTIONING ☐ NOTES

CALMNESS ☐ NOTES

CONCENTRATION ☐ NOTES

FOCUS / PRODUCTIVITY ☐ NOTES

ENERGY LEVELS ☐ NOTES

RELATIONAL SKILLS ☐ NOTES

SENSES ☐ NOTES

SLEEP ☐ NOTES

SENSE OF CONNECTION/ INTROSPECTION ☐ NOTES

DAILY BENEFIT SCORE ☐ NOTES

EFFECTS ON HEALTH ISSUES

NEGATIVE EFFECTS

SIDE EFFECTS (AND SEVERITY)

FEELINGS EOD

REFLECTION/OBSERVATIONS

DAY/TIME	DOSE	FORM	LAST INTAKE

GOALS/INTENTIONS

FEELINGS BOD

MOOD ☐ NOTES

COGNITIVE FUNCTIONING ☐ NOTES

CALMNESS ☐ NOTES

CONCENTRATION ☐ NOTES

FOCUS / PRODUCTIVITY ☐ NOTES

ENERGY LEVELS ☐ NOTES

RELATIONAL SKILLS ☐ NOTES

SENSES ☐ NOTES

SLEEP ☐ NOTES

**SENSE OF CONNECTION/
INTROSPECTION** ☐ NOTES

DAILY BENEFIT SCORE ☐ NOTES

EFFECTS ON HEALTH ISSUES

NEGATIVE EFFECTS

SIDE EFFECTS (AND SEVERITY)

FEELINGS EOD

REFLECTION/OBSERVATIONS

DAY/TIME	DOSE	FORM	LAST INTAKE

GOALS/INTENTIONS

FEELINGS BOD

MOOD	☐	NOTES
COGNITIVE FUNCTIONING	☐	NOTES
CALMNESS	☐	NOTES
CONCENTRATION	☐	NOTES
FOCUS / PRODUCTIVITY	☐	NOTES
ENERGY LEVELS	☐	NOTES
RELATIONAL SKILLS	☐	NOTES
SENSES	☐	NOTES
SLEEP	☐	NOTES
SENSE OF CONNECTION/ INTROSPECTION	☐	NOTES
DAILY BENEFIT SCORE	☐	NOTES

EFFECTS ON HEALTH ISSUES

NEGATIVE EFFECTS

SIDE EFFECTS (AND SEVERITY)

FEELINGS EOD

REFLECTION/OBSERVATIONS

DAY/TIME	DOSE	FORM	LAST INTAKE

GOALS/INTENTIONS

FEELINGS BOD

MOOD ☐ NOTES

COGNITIVE FUNCTIONING ☐ NOTES

CALMNESS ☐ NOTES

CONCENTRATION ☐ NOTES

FOCUS / PRODUCTIVITY ☐ NOTES

ENERGY LEVELS ☐ NOTES

RELATIONAL SKILLS ☐ NOTES

SENSES ☐ NOTES

SLEEP ☐ NOTES

SENSE OF CONNECTION/
INTROSPECTION ☐ NOTES

DAILY BENEFIT SCORE ☐ NOTES

EFFECTS ON HEALTH ISSUES

NEGATIVE EFFECTS

SIDE EFFECTS (AND SEVERITY)

FEELINGS EOD

REFLECTION/OBSERVATIONS

DAY/TIME	DOSE	FORM	LAST INTAKE

GOALS/INTENTIONS

FEELINGS BOD

MOOD ☐ NOTES

COGNITIVE FUNCTIONING ☐ NOTES

CALMNESS ☐ NOTES

CONCENTRATION ☐ NOTES

FOCUS / PRODUCTIVITY ☐ NOTES

ENERGY LEVELS ☐ NOTES

RELATIONAL SKILLS ☐ NOTES

SENSES ☐ NOTES

SLEEP ☐ NOTES

SENSE OF CONNECTION/
INTROSPECTION ☐ NOTES

DAILY BENEFIT SCORE ☐ NOTES

EFFECTS ON HEALTH ISSUES

NEGATIVE EFFECTS

SIDE EFFECTS (AND SEVERITY)

FEELINGS EOD

REFLECTION/OBSERVATIONS

DAY/TIME	DOSE	FORM	LAST INTAKE

GOALS/INTENTIONS

FEELINGS BOD

MOOD ☐ NOTES

COGNITIVE FUNCTIONING ☐ NOTES

CALMNESS ☐ NOTES

CONCENTRATION ☐ NOTES

FOCUS / PRODUCTIVITY ☐ NOTES

ENERGY LEVELS ☐ NOTES

RELATIONAL SKILLS ☐ NOTES

SENSES ☐ NOTES

SLEEP ☐ NOTES

SENSE OF CONNECTION/ INTROSPECTION ☐ NOTES

DAILY BENEFIT SCORE ☐ NOTES

EFFECTS ON HEALTH ISSUES

NEGATIVE EFFECTS

SIDE EFFECTS (AND SEVERITY)

FEELINGS EOD

REFLECTION/OBSERVATIONS

DAY/TIME	DOSE	FORM	LAST INTAKE

GOALS/INTENTIONS

FEELINGS BOD

MOOD ☐ NOTES

COGNITIVE FUNCTIONING ☐ NOTES

CALMNESS ☐ NOTES

CONCENTRATION ☐ NOTES

FOCUS / PRODUCTIVITY ☐ NOTES

ENERGY LEVELS ☐ NOTES

RELATIONAL SKILLS ☐ NOTES

SENSES ☐ NOTES

SLEEP ☐ NOTES

SENSE OF CONNECTION/ INTROSPECTION ☐ NOTES

DAILY BENEFIT SCORE ☐ NOTES

EFFECTS ON HEALTH ISSUES

NEGATIVE EFFECTS

SIDE EFFECTS (AND SEVERITY)

FEELINGS EOD

REFLECTION/OBSERVATIONS

DAY/TIME	DOSE	FORM	LAST INTAKE

GOALS/INTENTIONS

FEELINGS BOD

MOOD ☐ NOTES

COGNITIVE FUNCTIONING ☐ NOTES

CALMNESS ☐ NOTES

CONCENTRATION ☐ NOTES

FOCUS / PRODUCTIVITY ☐ NOTES

ENERGY LEVELS ☐ NOTES

RELATIONAL SKILLS ☐ NOTES

SENSES ☐ NOTES

SLEEP ☐ NOTES

SENSE OF CONNECTION/
INTROSPECTION ☐ NOTES

DAILY BENEFIT SCORE ☐ NOTES

EFFECTS ON HEALTH ISSUES

NEGATIVE EFFECTS

SIDE EFFECTS (AND SEVERITY)

FEELINGS EOD

REFLECTION/OBSERVATIONS

DAY/TIME	DOSE	FORM	LAST INTAKE

GOALS/INTENTIONS

FEELINGS BOD

MOOD ☐ NOTES

COGNITIVE FUNCTIONING ☐ NOTES

CALMNESS ☐ NOTES

CONCENTRATION ☐ NOTES

FOCUS / PRODUCTIVITY ☐ NOTES

ENERGY LEVELS ☐ NOTES

RELATIONAL SKILLS ☐ NOTES

SENSES ☐ NOTES

SLEEP ☐ NOTES

SENSE OF CONNECTION/ INTROSPECTION ☐ NOTES

DAILY BENEFIT SCORE ☐ NOTES

EFFECTS ON HEALTH ISSUES

NEGATIVE EFFECTS

SIDE EFFECTS (AND SEVERITY)

FEELINGS EOD

REFLECTION/OBSERVATIONS

DAY/TIME	DOSE	FORM	LAST INTAKE

GOALS/INTENTIONS

FEELINGS BOD

MOOD ☐ NOTES

COGNITIVE FUNCTIONING ☐ NOTES

CALMNESS ☐ NOTES

CONCENTRATION ☐ NOTES

FOCUS / PRODUCTIVITY ☐ NOTES

ENERGY LEVELS ☐ NOTES

RELATIONAL SKILLS ☐ NOTES

SENSES ☐ NOTES

SLEEP ☐ NOTES

SENSE OF CONNECTION/
INTROSPECTION ☐ NOTES

DAILY BENEFIT SCORE ☐ NOTES

EFFECTS ON HEALTH ISSUES

NEGATIVE EFFECTS

SIDE EFFECTS (AND SEVERITY)

FEELINGS EOD

REFLECTION/OBSERVATIONS

DAY/TIME	DOSE	FORM	LAST INTAKE

GOALS/INTENTIONS

FEELINGS BOD

MOOD ☐ NOTES

COGNITIVE FUNCTIONING ☐ NOTES

CALMNESS ☐ NOTES

CONCENTRATION ☐ NOTES

FOCUS / PRODUCTIVITY ☐ NOTES

ENERGY LEVELS ☐ NOTES

RELATIONAL SKILLS ☐ NOTES

SENSES ☐ NOTES

SLEEP ☐ NOTES

SENSE OF CONNECTION/ INTROSPECTION ☐ NOTES

DAILY BENEFIT SCORE ☐ NOTES

EFFECTS ON HEALTH ISSUES

NEGATIVE EFFECTS

SIDE EFFECTS (AND SEVERITY)

FEELINGS EOD

REFLECTION/OBSERVATIONS

DAY/TIME	DOSE	FORM	LAST INTAKE

GOALS/INTENTIONS

FEELINGS BOD

MOOD ☐ NOTES

COGNITIVE FUNCTIONING ☐ NOTES

CALMNESS ☐ NOTES

CONCENTRATION ☐ NOTES

FOCUS / PRODUCTIVITY ☐ NOTES

ENERGY LEVELS ☐ NOTES

RELATIONAL SKILLS ☐ NOTES

SENSES ☐ NOTES

SLEEP ☐ NOTES

SENSE OF CONNECTION/
INTROSPECTION ☐ NOTES

DAILY BENEFIT SCORE ☐ NOTES

EFFECTS ON HEALTH ISSUES

NEGATIVE EFFECTS

SIDE EFFECTS (AND SEVERITY)

FEELINGS EOD

REFLECTION/OBSERVATIONS

Monthly Overview

Overall Benefits Analyzer

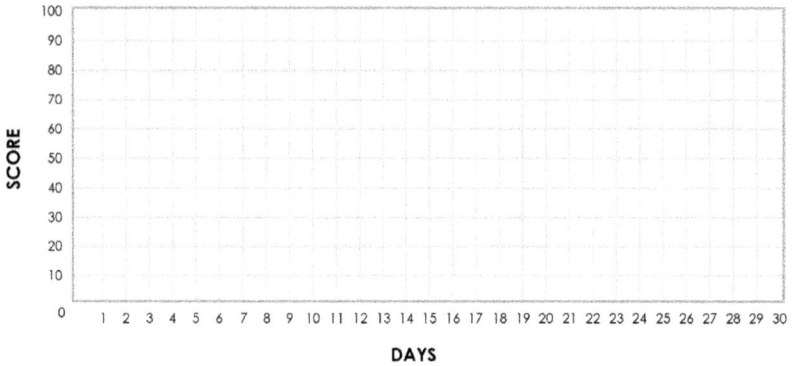

—————— Analyzer

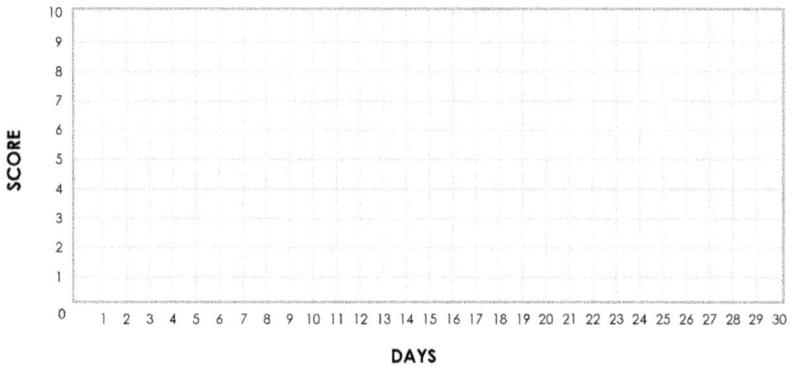

—————— Analyzer

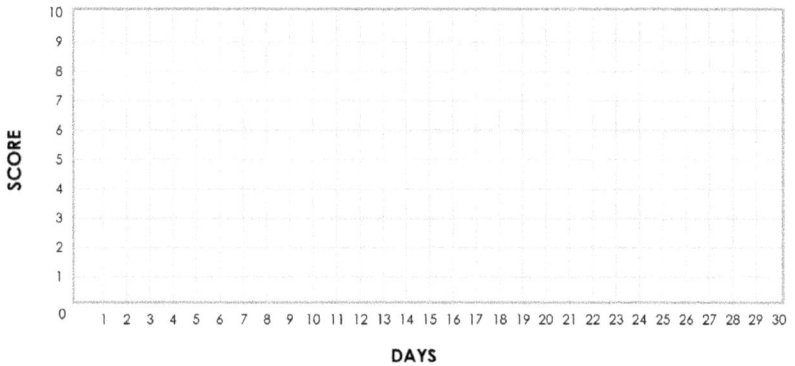

Monthly Overview

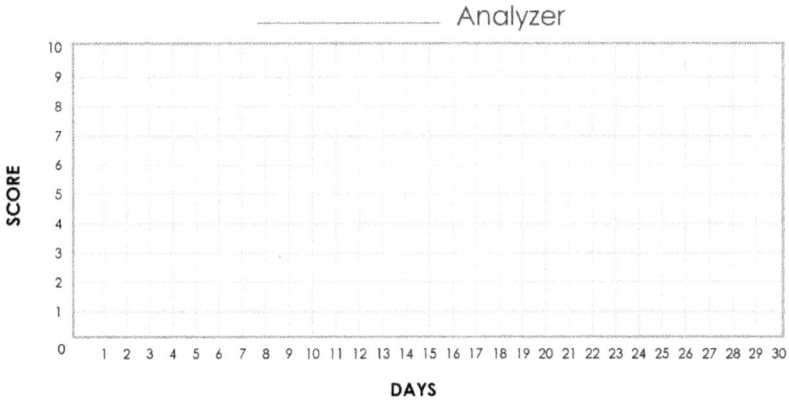

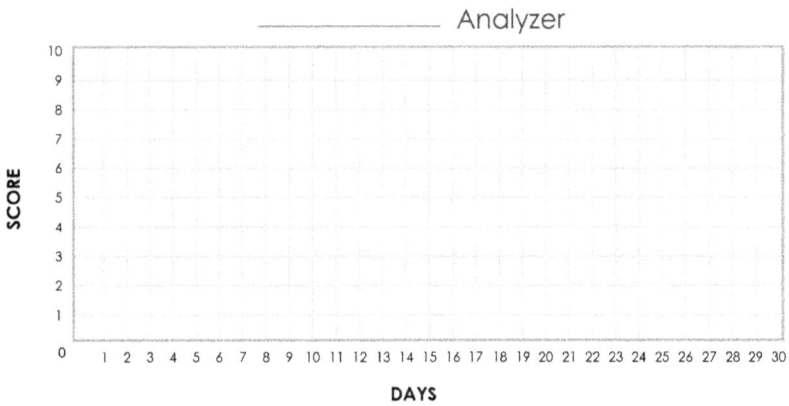

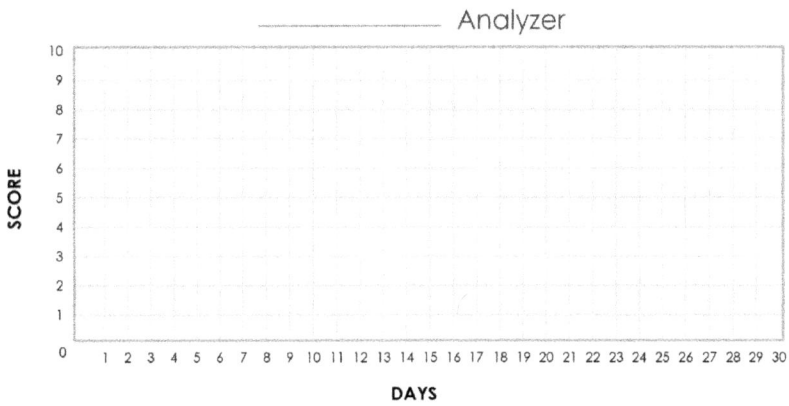

_____ Analyzer

SCORE

10
9
8
7
6
5
4
3
2
1
0

1 2 3 4 5 6 7 8 9 10 11 12 13 14 15 16 17 18 19 20 21 22 23 24 25 26 27 28 29 30

DAYS

_____ Analyzer

SCORE

10
9
8
7
6
5
4
3
2
1
0

1 2 3 4 5 6 7 8 9 10 11 12 13 14 15 16 17 18 19 20 21 22 23 24 25 26 27 28 29 30

DAYS

_____ Analyzer

SCORE

10
9
8
7
6
5
4
3
2
1
0

1 2 3 4 5 6 7 8 9 10 11 12 13 14 15 16 17 18 19 20 21 22 23 24 25 26 27 28 29 30

DAYS

Monthly Overview

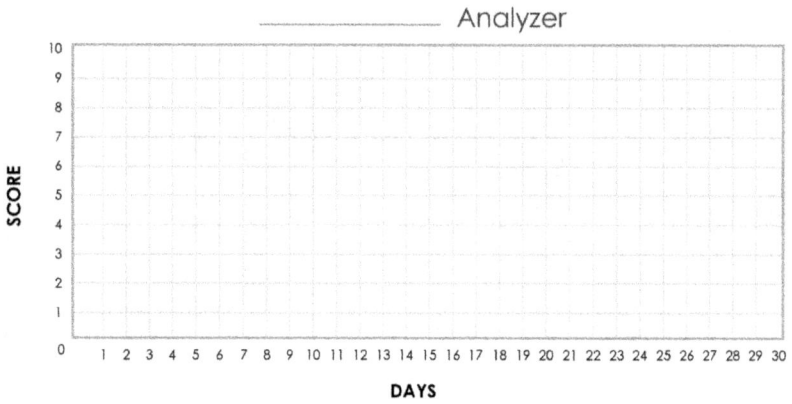

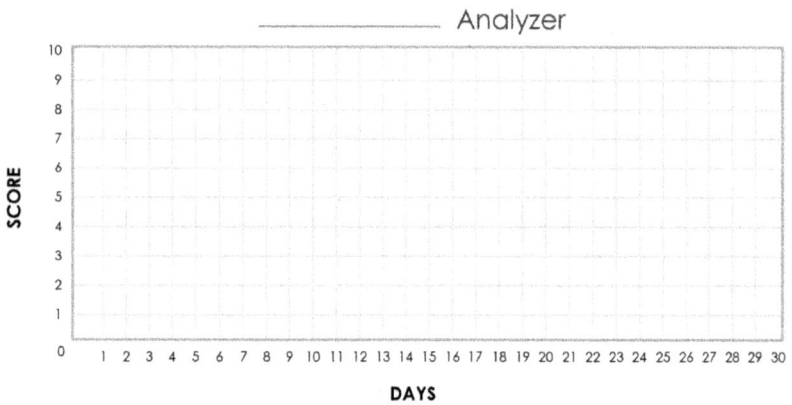

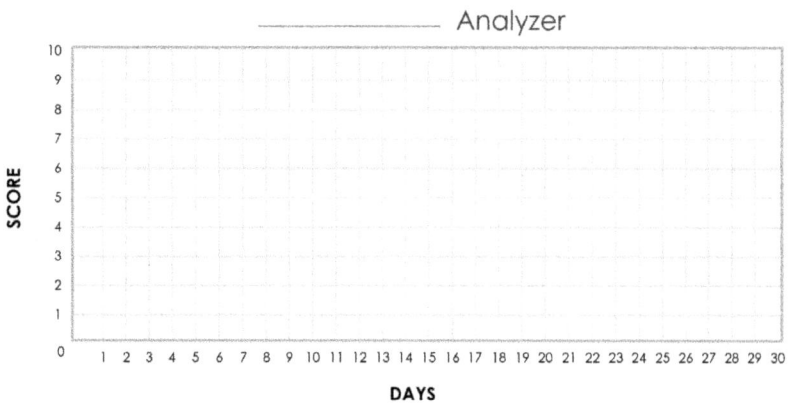

_____ Analyzer

SCORE

10
9
8
7
6
5
4
3
2
1
0

1 2 3 4 5 6 7 8 9 10 11 12 13 14 15 16 17 18 19 20 21 22 23 24 25 26 27 28 29 30

DAYS

_____ Analyzer

SCORE

10
9
8
7
6
5
4
3
2
1
0

1 2 3 4 5 6 7 8 9 10 11 12 13 14 15 16 17 18 19 20 21 22 23 24 25 26 27 28 29 30

DAYS

_____ Analyzer

SCORE

10
9
8
7
6
5
4
3
2
1
0

1 2 3 4 5 6 7 8 9 10 11 12 13 14 15 16 17 18 19 20 21 22 23 24 25 26 27 28 29 30

DAYS

Monthly Overview

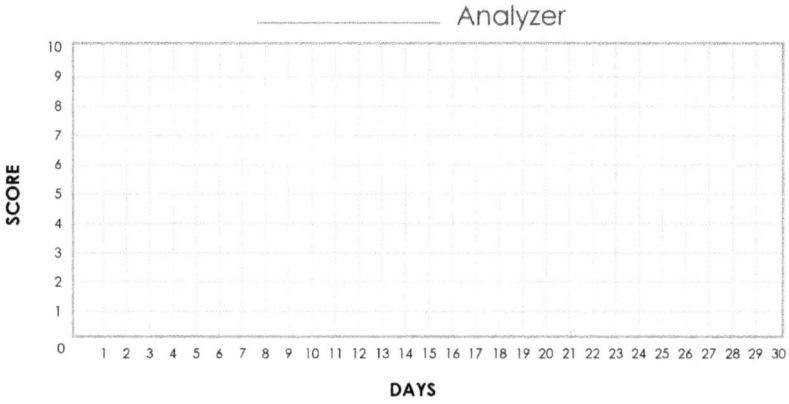

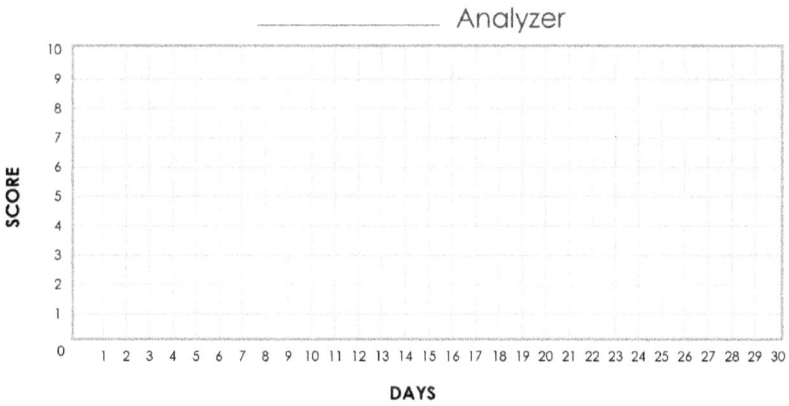

_____ Analyzer

SCORE

10
9
8
7
6
5
4
3
2
1
0

1 2 3 4 5 6 7 8 9 10 11 12 13 14 15 16 17 18 19 20 21 22 23 24 25 26 27 28 29 30

DAYS

_____ Analyzer

SCORE

10
9
8
7
6
5
4
3
2
1
0

1 2 3 4 5 6 7 8 9 10 11 12 13 14 15 16 17 18 19 20 21 22 23 24 25 26 27 28 29 30

DAYS

DAY/TIME	DOSE	FORM	LAST INTAKE

GOALS/INTENTIONS

FEELINGS BOD

MOOD ☐ NOTES

COGNITIVE FUNCTIONING ☐ NOTES

CALMNESS ☐ NOTES

CONCENTRATION ☐ NOTES

FOCUS / PRODUCTIVITY ☐ NOTES

ENERGY LEVELS ☐ NOTES

RELATIONAL SKILLS ☐ NOTES

SENSES ☐ NOTES

SLEEP ☐ NOTES

SENSE OF CONNECTION/
INTROSPECTION ☐ NOTES

DAILY BENEFIT SCORE ☐ NOTES

EFFECTS ON HEALTH ISSUES

NEGATIVE EFFECTS

SIDE EFFECTS (AND SEVERITY)

FEELINGS EOD

REFLECTION/OBSERVATIONS

DAY/TIME	DOSE	FORM	LAST INTAKE

GOALS/INTENTIONS

FEELINGS BOD

MOOD ☐ NOTES

COGNITIVE FUNCTIONING ☐ NOTES

CALMNESS ☐ NOTES

CONCENTRATION ☐ NOTES

FOCUS / PRODUCTIVITY ☐ NOTES

ENERGY LEVELS ☐ NOTES

RELATIONAL SKILLS ☐ NOTES

SENSES ☐ NOTES

SLEEP ☐ NOTES

SENSE OF CONNECTION/ INTROSPECTION ☐ NOTES

DAILY BENEFIT SCORE ☐ NOTES

EFFECTS ON HEALTH ISSUES

NEGATIVE EFFECTS

SIDE EFFECTS (AND SEVERITY)

FEELINGS EOD

REFLECTION/OBSERVATIONS

DAY/TIME	DOSE	FORM	LAST INTAKE

GOALS/INTENTIONS

FEELINGS BOD

MOOD ☐ NOTES

COGNITIVE FUNCTIONING ☐ NOTES

CALMNESS ☐ NOTES

CONCENTRATION ☐ NOTES

FOCUS / PRODUCTIVITY ☐ NOTES

ENERGY LEVELS ☐ NOTES

RELATIONAL SKILLS ☐ NOTES

SENSES ☐ NOTES

SLEEP ☐ NOTES

SENSE OF CONNECTION/ INTROSPECTION ☐ NOTES

DAILY BENEFIT SCORE ☐ NOTES

EFFECTS ON HEALTH ISSUES

NEGATIVE EFFECTS

SIDE EFFECTS (AND SEVERITY)

FEELINGS EOD

REFLECTION/OBSERVATIONS

DAY/TIME	DOSE	FORM	LAST INTAKE

GOALS/INTENTIONS

FEELINGS BOD

MOOD ☐ NOTES

COGNITIVE FUNCTIONING ☐ NOTES

CALMNESS ☐ NOTES

CONCENTRATION ☐ NOTES

FOCUS / PRODUCTIVITY ☐ NOTES

ENERGY LEVELS ☐ NOTES

RELATIONAL SKILLS ☐ NOTES

SENSES ☐ NOTES

SLEEP ☐ NOTES

SENSE OF CONNECTION/ INTROSPECTION ☐ NOTES

DAILY BENEFIT SCORE ☐ NOTES

EFFECTS ON HEALTH ISSUES

NEGATIVE EFFECTS

SIDE EFFECTS (AND SEVERITY)

FEELINGS EOD

REFLECTION/OBSERVATIONS

DAY/TIME	DOSE	FORM	LAST INTAKE

GOALS/INTENTIONS

FEELINGS BOD

MOOD ☐ NOTES

COGNITIVE FUNCTIONING ☐ NOTES

CALMNESS ☐ NOTES

CONCENTRATION ☐ NOTES

FOCUS / PRODUCTIVITY ☐ NOTES

ENERGY LEVELS ☐ NOTES

RELATIONAL SKILLS ☐ NOTES

SENSES ☐ NOTES

SLEEP ☐ NOTES

SENSE OF CONNECTION/
INTROSPECTION ☐ NOTES

DAILY BENEFIT SCORE ☐ NOTES

EFFECTS ON HEALTH ISSUES

NEGATIVE EFFECTS

SIDE EFFECTS (AND SEVERITY)

FEELINGS EOD

REFLECTION/OBSERVATIONS

DAY/TIME	DOSE	FORM	LAST INTAKE

GOALS/INTENTIONS

FEELINGS BOD

MOOD	☐	NOTES
COGNITIVE FUNCTIONING	☐	NOTES
CALMNESS	☐	NOTES
CONCENTRATION	☐	NOTES
FOCUS / PRODUCTIVITY	☐	NOTES
ENERGY LEVELS	☐	NOTES
RELATIONAL SKILLS	☐	NOTES
SENSES	☐	NOTES
SLEEP	☐	NOTES
SENSE OF CONNECTION/ INTROSPECTION	☐	NOTES
DAILY BENEFIT SCORE	☐	NOTES

EFFECTS ON HEALTH ISSUES

NEGATIVE EFFECTS

SIDE EFFECTS (AND SEVERITY)

FEELINGS EOD

REFLECTION/OBSERVATIONS

DAY/TIME	DOSE	FORM	LAST INTAKE

GOALS/INTENTIONS

FEELINGS BOD

MOOD ☐ NOTES

COGNITIVE FUNCTIONING ☐ NOTES

CALMNESS ☐ NOTES

CONCENTRATION ☐ NOTES

FOCUS / PRODUCTIVITY ☐ NOTES

ENERGY LEVELS ☐ NOTES

RELATIONAL SKILLS ☐ NOTES

SENSES ☐ NOTES

SLEEP ☐ NOTES

SENSE OF CONNECTION/
INTROSPECTION ☐ NOTES

DAILY BENEFIT SCORE ☐ NOTES

EFFECTS ON HEALTH ISSUES

NEGATIVE EFFECTS

SIDE EFFECTS (AND SEVERITY)

FEELINGS EOD

REFLECTION/OBSERVATIONS

DAY/TIME	DOSE	FORM	LAST INTAKE

GOALS/INTENTIONS

FEELINGS BOD

MOOD ☐ NOTES

COGNITIVE FUNCTIONING ☐ NOTES

CALMNESS ☐ NOTES

CONCENTRATION ☐ NOTES

FOCUS / PRODUCTIVITY ☐ NOTES

ENERGY LEVELS ☐ NOTES

RELATIONAL SKILLS ☐ NOTES

SENSES ☐ NOTES

SLEEP ☐ NOTES

SENSE OF CONNECTION/ INTROSPECTION ☐ NOTES

DAILY BENEFIT SCORE ☐ NOTES

EFFECTS ON HEALTH ISSUES

NEGATIVE EFFECTS

SIDE EFFECTS (AND SEVERITY)

FEELINGS EOD

REFLECTION/OBSERVATIONS

DAY/TIME	DOSE	FORM	LAST INTAKE

GOALS/INTENTIONS

FEELINGS BOD

MOOD ☐ NOTES

COGNITIVE FUNCTIONING ☐ NOTES

CALMNESS ☐ NOTES

CONCENTRATION ☐ NOTES

FOCUS / PRODUCTIVITY ☐ NOTES

ENERGY LEVELS ☐ NOTES

RELATIONAL SKILLS ☐ NOTES

SENSES ☐ NOTES

SLEEP ☐ NOTES

SENSE OF CONNECTION/ INTROSPECTION ☐ NOTES

DAILY BENEFIT SCORE ☐ NOTES

EFFECTS ON HEALTH ISSUES

NEGATIVE EFFECTS

SIDE EFFECTS (AND SEVERITY)

FEELINGS EOD

REFLECTION/OBSERVATIONS

DAY/TIME	DOSE	FORM	LAST INTAKE

GOALS/INTENTIONS

FEELINGS BOD

MOOD ☐ NOTES

COGNITIVE FUNCTIONING ☐ NOTES

CALMNESS ☐ NOTES

CONCENTRATION ☐ NOTES

FOCUS / PRODUCTIVITY ☐ NOTES

ENERGY LEVELS ☐ NOTES

RELATIONAL SKILLS ☐ NOTES

SENSES ☐ NOTES

SLEEP ☐ NOTES

SENSE OF CONNECTION/ INTROSPECTION ☐ NOTES

DAILY BENEFIT SCORE ☐ NOTES

EFFECTS ON HEALTH ISSUES

NEGATIVE EFFECTS

SIDE EFFECTS (AND SEVERITY)

FEELINGS EOD

REFLECTION/OBSERVATIONS

DAY/TIME	DOSE	FORM	LAST INTAKE

GOALS/INTENTIONS

FEELINGS BOD

MOOD	☐	NOTES
COGNITIVE FUNCTIONING	☐	NOTES
CALMNESS	☐	NOTES
CONCENTRATION	☐	NOTES
FOCUS / PRODUCTIVITY	☐	NOTES
ENERGY LEVELS	☐	NOTES
RELATIONAL SKILLS	☐	NOTES
SENSES	☐	NOTES
SLEEP	☐	NOTES
SENSE OF CONNECTION/ INTROSPECTION	☐	NOTES
DAILY BENEFIT SCORE	☐	NOTES

EFFECTS ON HEALTH ISSUES

NEGATIVE EFFECTS

SIDE EFFECTS (AND SEVERITY)

FEELINGS EOD

REFLECTION/OBSERVATIONS

DAY/TIME	DOSE	FORM	LAST INTAKE

GOALS/INTENTIONS

FEELINGS BOD

MOOD	☐	NOTES
COGNITIVE FUNCTIONING	☐	NOTES
CALMNESS	☐	NOTES
CONCENTRATION	☐	NOTES
FOCUS / PRODUCTIVITY	☐	NOTES
ENERGY LEVELS	☐	NOTES
RELATIONAL SKILLS	☐	NOTES
SENSES	☐	NOTES
SLEEP	☐	NOTES
SENSE OF CONNECTION/ INTROSPECTION	☐	NOTES
DAILY BENEFIT SCORE	☐	NOTES

EFFECTS ON HEALTH ISSUES

NEGATIVE EFFECTS

SIDE EFFECTS (AND SEVERITY)

FEELINGS EOD

REFLECTION/OBSERVATIONS

DAY/TIME	DOSE	FORM	LAST INTAKE

GOALS/INTENTIONS

FEELINGS BOD

MOOD ☐ NOTES

COGNITIVE FUNCTIONING ☐ NOTES

CALMNESS ☐ NOTES

CONCENTRATION ☐ NOTES

FOCUS / PRODUCTIVITY ☐ NOTES

ENERGY LEVELS ☐ NOTES

RELATIONAL SKILLS ☐ NOTES

SENSES ☐ NOTES

SLEEP ☐ NOTES

SENSE OF CONNECTION/ INTROSPECTION ☐ NOTES

DAILY BENEFIT SCORE ☐ NOTES

EFFECTS ON HEALTH ISSUES

NEGATIVE EFFECTS

SIDE EFFECTS (AND SEVERITY)

FEELINGS EOD

REFLECTION/OBSERVATIONS

DAY/TIME	DOSE	FORM	LAST INTAKE

GOALS/INTENTIONS

FEELINGS BOD

MOOD ☐ NOTES

COGNITIVE FUNCTIONING ☐ NOTES

CALMNESS ☐ NOTES

CONCENTRATION ☐ NOTES

FOCUS / PRODUCTIVITY ☐ NOTES

ENERGY LEVELS ☐ NOTES

RELATIONAL SKILLS ☐ NOTES

SENSES ☐ NOTES

SLEEP ☐ NOTES

SENSE OF CONNECTION/ INTROSPECTION ☐ NOTES

DAILY BENEFIT SCORE ☐ NOTES

EFFECTS ON HEALTH ISSUES

NEGATIVE EFFECTS

SIDE EFFECTS (AND SEVERITY)

FEELINGS EOD

REFLECTION/OBSERVATIONS

DAY/TIME	DOSE	FORM	LAST INTAKE

GOALS/INTENTIONS

FEELINGS BOD

MOOD	☐	NOTES
COGNITIVE FUNCTIONING	☐	NOTES
CALMNESS	☐	NOTES
CONCENTRATION	☐	NOTES
FOCUS / PRODUCTIVITY	☐	NOTES
ENERGY LEVELS	☐	NOTES
RELATIONAL SKILLS	☐	NOTES
SENSES	☐	NOTES
SLEEP	☐	NOTES
SENSE OF CONNECTION/ INTROSPECTION	☐	NOTES
DAILY BENEFIT SCORE	☐	NOTES

EFFECTS ON HEALTH ISSUES

NEGATIVE EFFECTS

SIDE EFFECTS (AND SEVERITY)

FEELINGS EOD

REFLECTION/OBSERVATIONS

DAY/TIME	DOSE	FORM	LAST INTAKE

GOALS/INTENTIONS

FEELINGS BOD

MOOD ☐ NOTES

COGNITIVE FUNCTIONING ☐ NOTES

CALMNESS ☐ NOTES

CONCENTRATION ☐ NOTES

FOCUS / PRODUCTIVITY ☐ NOTES

ENERGY LEVELS ☐ NOTES

RELATIONAL SKILLS ☐ NOTES

SENSES ☐ NOTES

SLEEP ☐ NOTES

SENSE OF CONNECTION/
INTROSPECTION ☐ NOTES

DAILY BENEFIT SCORE ☐ NOTES

EFFECTS ON HEALTH ISSUES

NEGATIVE EFFECTS

SIDE EFFECTS (AND SEVERITY)

FEELINGS EOD

REFLECTION/OBSERVATIONS

DAY/TIME	DOSE	FORM	LAST INTAKE

GOALS/INTENTIONS

FEELINGS BOD

MOOD ☐ NOTES

COGNITIVE FUNCTIONING ☐ NOTES

CALMNESS ☐ NOTES

CONCENTRATION ☐ NOTES

FOCUS / PRODUCTIVITY ☐ NOTES

ENERGY LEVELS ☐ NOTES

RELATIONAL SKILLS ☐ NOTES

SENSES ☐ NOTES

SLEEP ☐ NOTES

SENSE OF CONNECTION/ INTROSPECTION ☐ NOTES

DAILY BENEFIT SCORE ☐ NOTES

EFFECTS ON HEALTH ISSUES

NEGATIVE EFFECTS

SIDE EFFECTS (AND SEVERITY)

FEELINGS EOD

REFLECTION/OBSERVATIONS

DAY/TIME	DOSE	FORM	LAST INTAKE

GOALS/INTENTIONS

FEELINGS BOD

MOOD ☐ NOTES

COGNITIVE FUNCTIONING ☐ NOTES

CALMNESS ☐ NOTES

CONCENTRATION ☐ NOTES

FOCUS / PRODUCTIVITY ☐ NOTES

ENERGY LEVELS ☐ NOTES

RELATIONAL SKILLS ☐ NOTES

SENSES ☐ NOTES

SLEEP ☐ NOTES

SENSE OF CONNECTION/ INTROSPECTION ☐ NOTES

DAILY BENEFIT SCORE ☐ NOTES

EFFECTS ON HEALTH ISSUES

NEGATIVE EFFECTS

SIDE EFFECTS (AND SEVERITY)

FEELINGS EOD

REFLECTION/OBSERVATIONS

DAY/TIME	DOSE	FORM	LAST INTAKE

GOALS/INTENTIONS

FEELINGS BOD

MOOD	☐	NOTES
COGNITIVE FUNCTIONING	☐	NOTES
CALMNESS	☐	NOTES
CONCENTRATION	☐	NOTES
FOCUS / PRODUCTIVITY	☐	NOTES
ENERGY LEVELS	☐	NOTES
RELATIONAL SKILLS	☐	NOTES
SENSES	☐	NOTES
SLEEP	☐	NOTES
SENSE OF CONNECTION/ INTROSPECTION	☐	NOTES
DAILY BENEFIT SCORE	☐	NOTES

EFFECTS ON HEALTH ISSUES

NEGATIVE EFFECTS

SIDE EFFECTS (AND SEVERITY)

FEELINGS EOD

REFLECTION/OBSERVATIONS

DAY/TIME	DOSE	FORM	LAST INTAKE

GOALS/INTENTIONS

FEELINGS BOD

MOOD ☐ NOTES

COGNITIVE FUNCTIONING ☐ NOTES

CALMNESS ☐ NOTES

CONCENTRATION ☐ NOTES

FOCUS / PRODUCTIVITY ☐ NOTES

ENERGY LEVELS ☐ NOTES

RELATIONAL SKILLS ☐ NOTES

SENSES ☐ NOTES

SLEEP ☐ NOTES

SENSE OF CONNECTION/
INTROSPECTION ☐ NOTES

DAILY BENEFIT SCORE ☐ NOTES

EFFECTS ON HEALTH ISSUES

NEGATIVE EFFECTS

SIDE EFFECTS (AND SEVERITY)

FEELINGS EOD

REFLECTION/OBSERVATIONS

DAY/TIME	DOSE	FORM	LAST INTAKE

GOALS/INTENTIONS

FEELINGS BOD

MOOD ☐ NOTES

COGNITIVE FUNCTIONING ☐ NOTES

CALMNESS ☐ NOTES

CONCENTRATION ☐ NOTES

FOCUS / PRODUCTIVITY ☐ NOTES

ENERGY LEVELS ☐ NOTES

RELATIONAL SKILLS ☐ NOTES

SENSES ☐ NOTES

SLEEP ☐ NOTES

SENSE OF CONNECTION/ INTROSPECTION ☐ NOTES

DAILY BENEFIT SCORE ☐ NOTES

EFFECTS ON HEALTH ISSUES

NEGATIVE EFFECTS

SIDE EFFECTS (AND SEVERITY)

FEELINGS EOD

REFLECTION/OBSERVATIONS

DAY/TIME	DOSE	FORM	LAST INTAKE

GOALS/INTENTIONS

FEELINGS BOD

MOOD ☐ NOTES

COGNITIVE FUNCTIONING ☐ NOTES

CALMNESS ☐ NOTES

CONCENTRATION ☐ NOTES

FOCUS / PRODUCTIVITY ☐ NOTES

ENERGY LEVELS ☐ NOTES

RELATIONAL SKILLS ☐ NOTES

SENSES ☐ NOTES

SLEEP ☐ NOTES

SENSE OF CONNECTION/ INTROSPECTION ☐ NOTES

DAILY BENEFIT SCORE ☐ NOTES

EFFECTS ON HEALTH ISSUES

NEGATIVE EFFECTS

SIDE EFFECTS (AND SEVERITY)

FEELINGS EOD

REFLECTION/OBSERVATIONS

DAY/TIME	DOSE	FORM	LAST INTAKE

GOALS/INTENTIONS

FEELINGS BOD

MOOD ☐ NOTES

COGNITIVE FUNCTIONING ☐ NOTES

CALMNESS ☐ NOTES

CONCENTRATION ☐ NOTES

FOCUS / PRODUCTIVITY ☐ NOTES

ENERGY LEVELS ☐ NOTES

RELATIONAL SKILLS ☐ NOTES

SENSES ☐ NOTES

SLEEP ☐ NOTES

SENSE OF CONNECTION/
INTROSPECTION ☐ NOTES

DAILY BENEFIT SCORE ☐ NOTES

EFFECTS ON HEALTH ISSUES

NEGATIVE EFFECTS

SIDE EFFECTS (AND SEVERITY)

FEELINGS EOD

REFLECTION/OBSERVATIONS

DAY/TIME	DOSE	FORM	LAST INTAKE

GOALS/INTENTIONS

FEELINGS BOD

MOOD ☐ NOTES

COGNITIVE FUNCTIONING ☐ NOTES

CALMNESS ☐ NOTES

CONCENTRATION ☐ NOTES

FOCUS / PRODUCTIVITY ☐ NOTES

ENERGY LEVELS ☐ NOTES

RELATIONAL SKILLS ☐ NOTES

SENSES ☐ NOTES

SLEEP ☐ NOTES

SENSE OF CONNECTION/ INTROSPECTION ☐ NOTES

DAILY BENEFIT SCORE ☐ NOTES

EFFECTS ON HEALTH ISSUES

NEGATIVE EFFECTS

SIDE EFFECTS (AND SEVERITY)

FEELINGS EOD

REFLECTION/OBSERVATIONS

DAY/TIME	DOSE	FORM	LAST INTAKE

GOALS/INTENTIONS

FEELINGS BOD

MOOD ☐ NOTES

COGNITIVE FUNCTIONING ☐ NOTES

CALMNESS ☐ NOTES

CONCENTRATION ☐ NOTES

FOCUS / PRODUCTIVITY ☐ NOTES

ENERGY LEVELS ☐ NOTES

RELATIONAL SKILLS ☐ NOTES

SENSES ☐ NOTES

SLEEP ☐ NOTES

SENSE OF CONNECTION/
INTROSPECTION ☐ NOTES

DAILY BENEFIT SCORE ☐ NOTES

EFFECTS ON HEALTH ISSUES

NEGATIVE EFFECTS

SIDE EFFECTS (AND SEVERITY)

FEELINGS EOD

REFLECTION/OBSERVATIONS

DAY/TIME	DOSE	FORM	LAST INTAKE

GOALS/INTENTIONS

FEELINGS BOD

MOOD ☐ NOTES

COGNITIVE FUNCTIONING ☐ NOTES

CALMNESS ☐ NOTES

CONCENTRATION ☐ NOTES

FOCUS / PRODUCTIVITY ☐ NOTES

ENERGY LEVELS ☐ NOTES

RELATIONAL SKILLS ☐ NOTES

SENSES ☐ NOTES

SLEEP ☐ NOTES

SENSE OF CONNECTION/ INTROSPECTION ☐ NOTES

DAILY BENEFIT SCORE ☐ NOTES

EFFECTS ON HEALTH ISSUES

NEGATIVE EFFECTS

SIDE EFFECTS (AND SEVERITY)

FEELINGS EOD

REFLECTION/OBSERVATIONS

DAY/TIME	DOSE	FORM	LAST INTAKE

GOALS/INTENTIONS

FEELINGS BOD

MOOD	☐	NOTES
COGNITIVE FUNCTIONING	☐	NOTES
CALMNESS	☐	NOTES
CONCENTRATION	☐	NOTES
FOCUS / PRODUCTIVITY	☐	NOTES
ENERGY LEVELS	☐	NOTES
RELATIONAL SKILLS	☐	NOTES
SENSES	☐	NOTES
SLEEP	☐	NOTES
SENSE OF CONNECTION/ INTROSPECTION	☐	NOTES
DAILY BENEFIT SCORE	☐	NOTES

EFFECTS ON HEALTH ISSUES

NEGATIVE EFFECTS

SIDE EFFECTS (AND SEVERITY)

FEELINGS EOD

REFLECTION/OBSERVATIONS

DAY/TIME	DOSE	FORM	LAST INTAKE

GOALS/INTENTIONS

FEELINGS BOD

MOOD ☐ NOTES

COGNITIVE FUNCTIONING ☐ NOTES

CALMNESS ☐ NOTES

CONCENTRATION ☐ NOTES

FOCUS / PRODUCTIVITY ☐ NOTES

ENERGY LEVELS ☐ NOTES

RELATIONAL SKILLS ☐ NOTES

SENSES ☐ NOTES

SLEEP ☐ NOTES

SENSE OF CONNECTION/
INTROSPECTION ☐ NOTES

DAILY BENEFIT SCORE ☐ NOTES

EFFECTS ON HEALTH ISSUES

NEGATIVE EFFECTS

SIDE EFFECTS (AND SEVERITY)

FEELINGS EOD

REFLECTION/OBSERVATIONS

DAY/TIME	DOSE	FORM	LAST INTAKE

GOALS/INTENTIONS

FEELINGS BOD

MOOD ☐ NOTES

COGNITIVE FUNCTIONING ☐ NOTES

CALMNESS ☐ NOTES

CONCENTRATION ☐ NOTES

FOCUS / PRODUCTIVITY ☐ NOTES

ENERGY LEVELS ☐ NOTES

RELATIONAL SKILLS ☐ NOTES

SENSES ☐ NOTES

SLEEP ☐ NOTES

SENSE OF CONNECTION/ INTROSPECTION ☐ NOTES

DAILY BENEFIT SCORE ☐ NOTES

EFFECTS ON HEALTH ISSUES

NEGATIVE EFFECTS

SIDE EFFECTS (AND SEVERITY)

FEELINGS EOD

REFLECTION/OBSERVATIONS

DAY/TIME	DOSE	FORM	LAST INTAKE

GOALS/INTENTIONS

FEELINGS BOD

MOOD	☐	NOTES
COGNITIVE FUNCTIONING	☐	NOTES
CALMNESS	☐	NOTES
CONCENTRATION	☐	NOTES
FOCUS / PRODUCTIVITY	☐	NOTES
ENERGY LEVELS	☐	NOTES
RELATIONAL SKILLS	☐	NOTES
SENSES	☐	NOTES
SLEEP	☐	NOTES
SENSE OF CONNECTION/ INTROSPECTION	☐	NOTES
DAILY BENEFIT SCORE	☐	NOTES

EFFECTS ON HEALTH ISSUES

NEGATIVE EFFECTS

SIDE EFFECTS (AND SEVERITY)

FEELINGS EOD

REFLECTION/OBSERVATIONS

Monthly Overview

Overall Benefits Analyzer

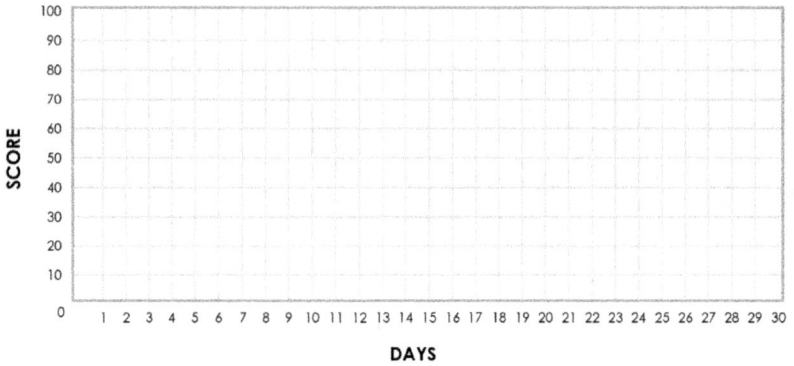

_____ Analyzer

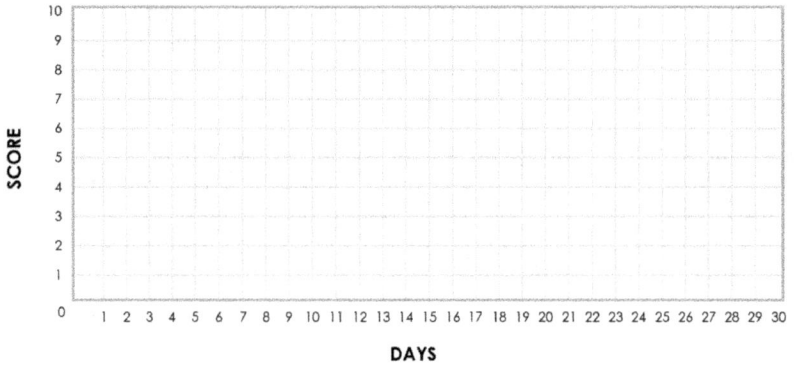

_____ Analyzer

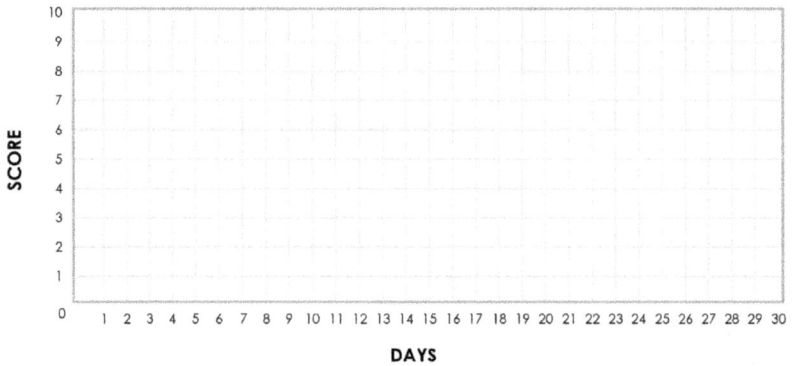

Monthly Overview

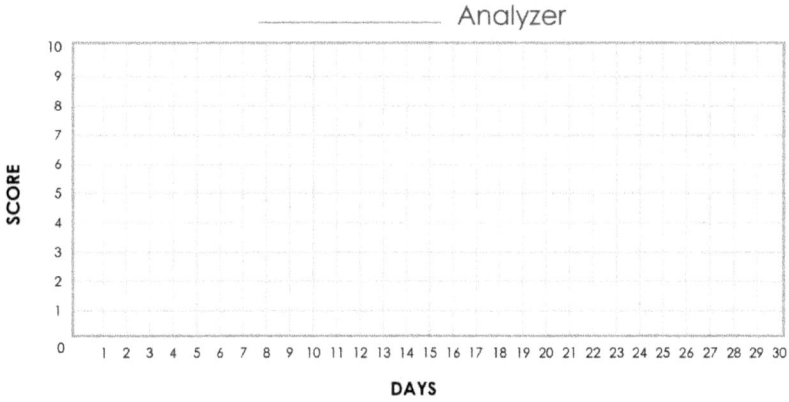

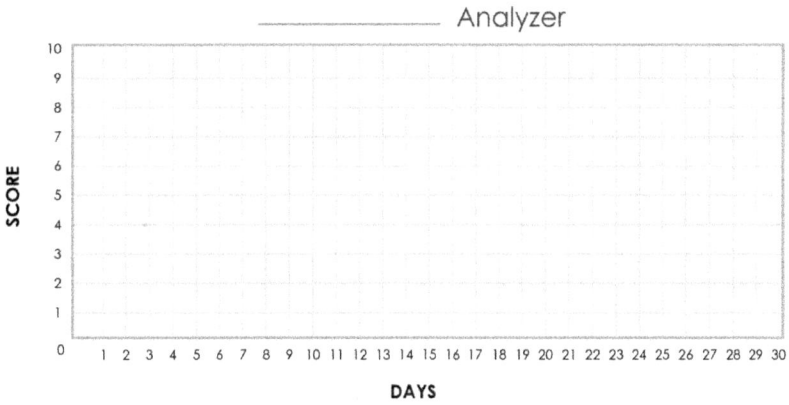

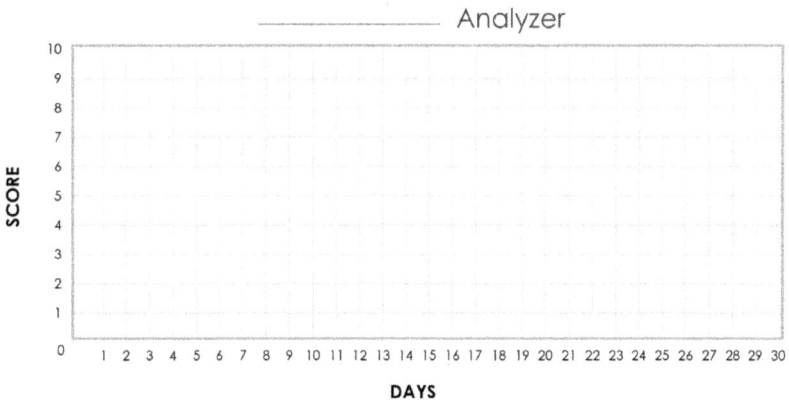

_____ Analyzer

SCORE

10
9
8
7
6
5
4
3
2
1
0

1 2 3 4 5 6 7 8 9 10 11 12 13 14 15 16 17 18 19 20 21 22 23 24 25 26 27 28 29 30

DAYS

_____ Analyzer

SCORE

10
9
8
7
6
5
4
3
2
1
0

1 2 3 4 5 6 7 8 9 10 11 12 13 14 15 16 17 18 19 20 21 22 23 24 25 26 27 28 29 30

DAYS

_____ Analyzer

SCORE

10
9
8
7
6
5
4
3
2
1
0

1 2 3 4 5 6 7 8 9 10 11 12 13 14 15 16 17 18 19 20 21 22 23 24 25 26 27 28 29 30

DAYS

Monthly Overview

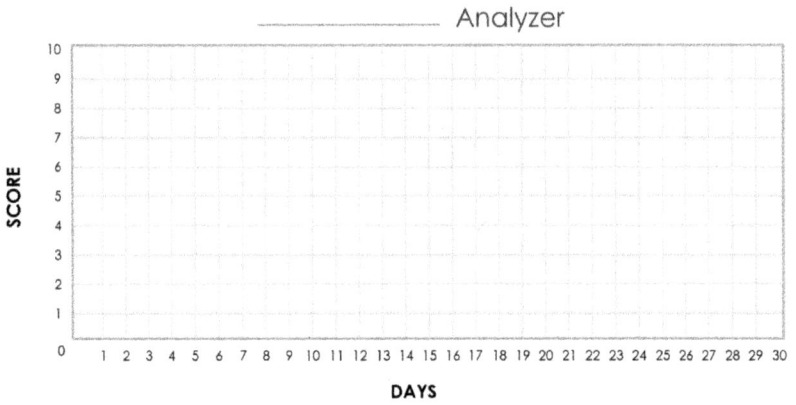

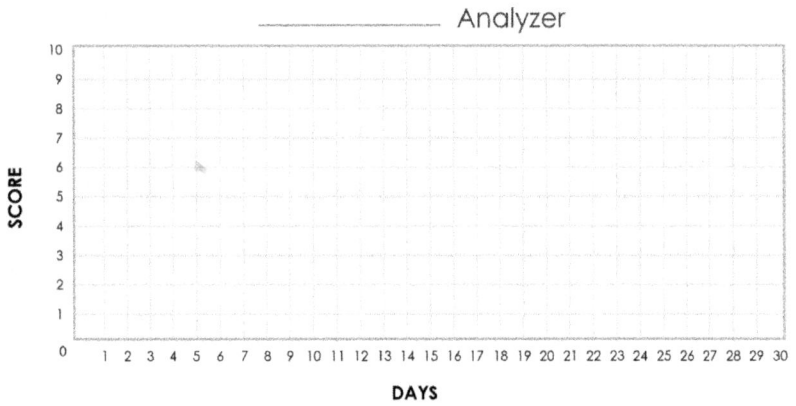

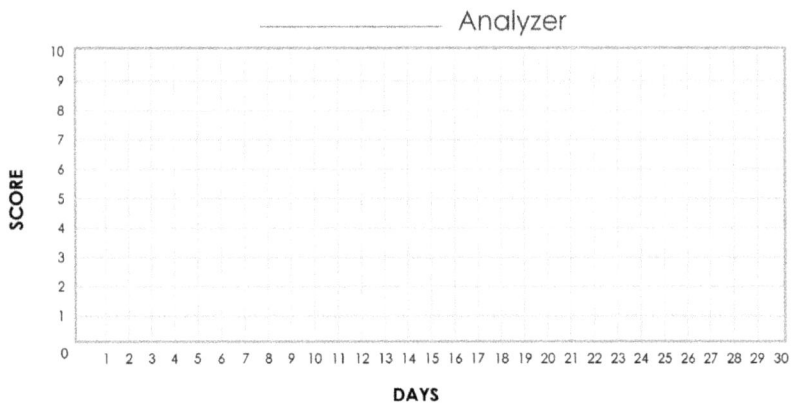

—————— Analyzer

SCORE	10 9 8 7 6 5 4 3 2 1 0

DAYS

1 2 3 4 5 6 7 8 9 10 11 12 13 14 15 16 17 18 19 20 21 22 23 24 25 26 27 28 29 30

—————— Analyzer

SCORE	10 9 8 7 6 5 4 3 2 1 0

DAYS

1 2 3 4 5 6 7 8 9 10 11 12 13 14 15 16 17 18 19 20 21 22 23 24 25 26 27 28 29 30

—————— Analyzer

SCORE	10 9 8 7 6 5 4 3 2 1 0

DAYS

1 2 3 4 5 6 7 8 9 10 11 12 13 14 15 16 17 18 19 20 21 22 23 24 25 26 27 28 29 30

Monthly Overview

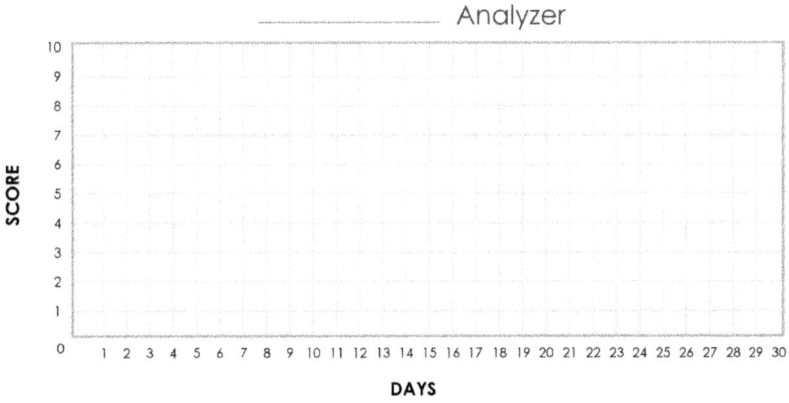

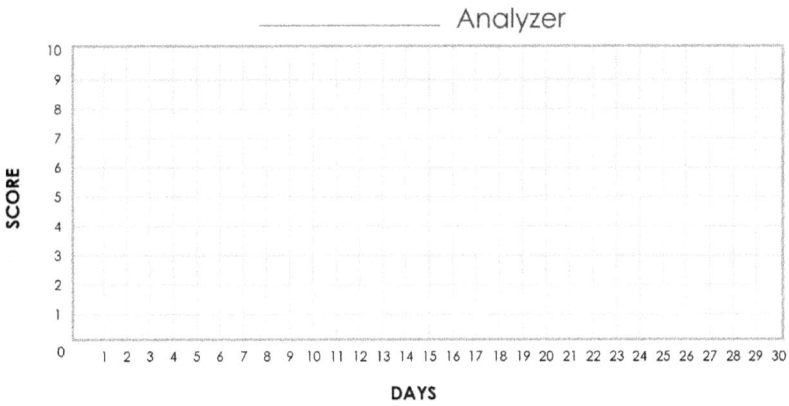

_____ Analyzer

SCORE

10
9
8
7
6
5
4
3
2
1
0

1 2 3 4 5 6 7 8 9 10 11 12 13 14 15 16 17 18 19 20 21 22 23 24 25 26 27 28 29 30

DAYS

_____ Analyzer

SCORE

10
9
8
7
6
5
4
3
2
1
0

1 2 3 4 5 6 7 8 9 10 11 12 13 14 15 16 17 18 19 20 21 22 23 24 25 26 27 28 29 30

DAYS

Dear Reader

I want to personally thank you for choosing this journal from among dozens out there, and for supporting my work.

If you enjoyed my work, please consider posting a review or rating on Amazon, it would mean a lot to meand help others benefit from it. It is also the best way to support independent writers like myself.

Thank you.

Leave a review Amazon US

Leave a review Amazon UK

You can use your respective Amazon market if you don't live in the UK or US.

www.ingramcontent.com/pod-product-compliance
Lightning Source LLC
Chambersburg PA
CBHW070109030426
42335CB00016B/2072